ISBN 978-1-330-37349-1
PIBN 10043288

This book is a reproduction of an important historical work. Forgotten Books uses
state-of-the-art technology to digitally reconstruct the work, preserving the original format
whilst repairing imperfections present in the aged copy. In rare cases, an imperfection in
the original, such as a blemish or missing page, may be replicated in our edition. We do,
however, repair the vast majority of imperfections successfully; any imperfections that
remain are intentionally left to preserve the state of such historical works.

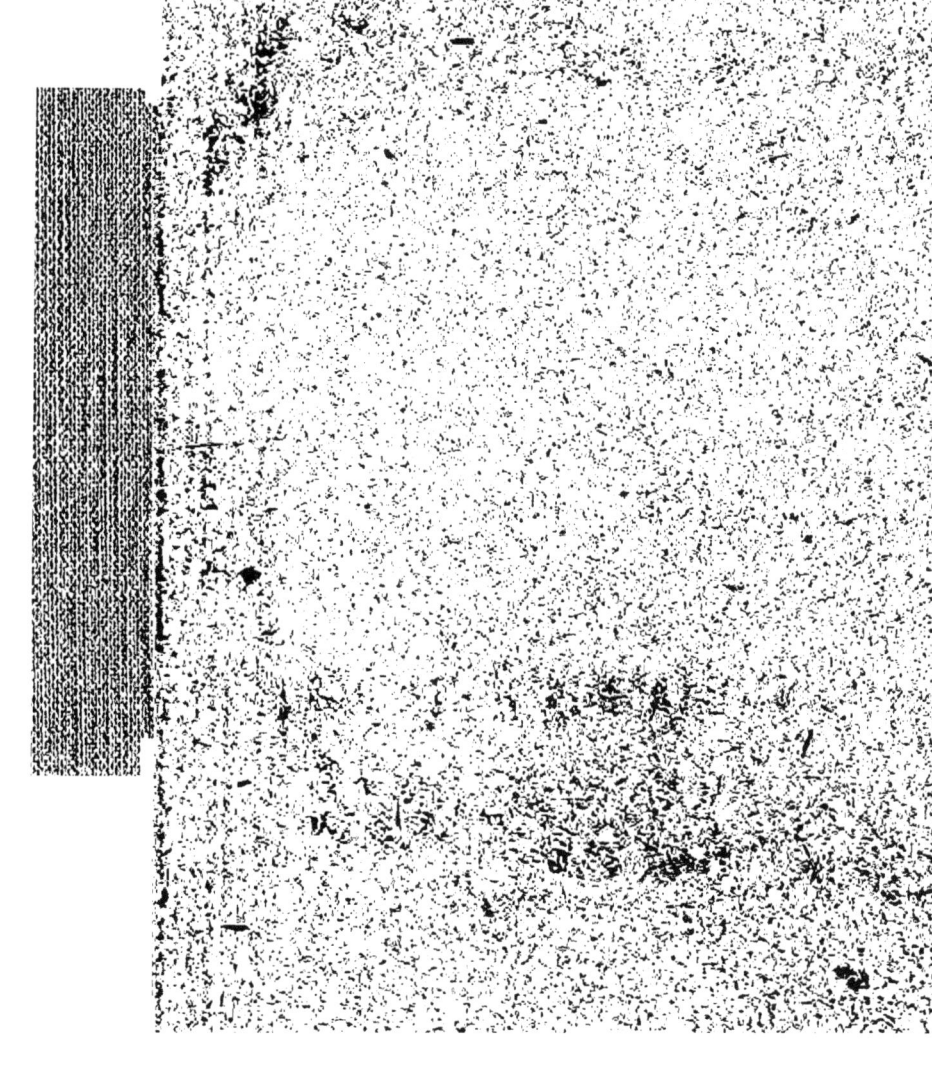

**THE
PEOPLE'S
BOOKS**

PHYSICAL CHEMISTRY

PHYSICAL CHEMISTRY

BY

William Cudmore McCullagh

W. C. McC. LEWIS, M.A., D.Sc.

BRUNNER PROFESSOR OF PHYSICAL CHEMISTRY IN THE
UNIVERSITY OF LIVERPOOL; AUTHOR OF "A SYSTEM
OF PHYSICAL CHEMISTRY," IN 3 VOLS.

London and Edinburgh:

T. C. & E. C. JACK, LTD. | T. NELSON & SONS, LTD.

1920

PREFACE.

In writing this little book it has been assumed that the reader is familiar with the elements of inorganic chemistry, and has likewise some slight knowledge of physics. Those who may not possess this are recommended, before taking up the subject of physical chemistry, to read the following books which have already appeared in this series.—

Inorganic Chemistry. By Professor E. C. C. Baly, F.R.S.

Organic Chemistry. By Professor J. B. Cohen, F.R.S.

Principles of Electricity. By Dr. N. Campbell.

I have to acknowledge the permission kindly granted me by the Editor of *Science Progress* to incorporate an article upon " The Structure of Matter " which appeared in that journal in 1918.

W. C. M'C. L.

CONTENTS.

I. Introductory 9

II. The Physical States of Matter 27

III. Molecular Magnitudes and the Structure of the Atom . . 58

IV. Chemical Reactions in Gases . 76

V. Chemical Reactions in Solutions 97

VI. Heats of Reactions . . . 119

PHYSICAL CHEMISTRY.

CHAPTER I.

INTRODUCTORY.

What is Physical Chemistry?—A definition which is sometimes given is as follows : " Physical chemistry consists of the application of physical methods to the elucidation of chemical problems."

This is certainly true as far as it goes, but as a comprehensive description it is wholly inadequate. Properly speaking, physical chemistry is *a new point of view* in the treatment of the problems of organic and inorganic chemistry. Physical chemistry deals essentially with the nature of the mechanism of chemical phenomena.

In the evolution of any science—the science of chemistry, for example—we naturally advance from the descriptive to the rational—*i.e.*, from the stage of observing chemical phenomena to the stage at which we attempt to account for and correlate them. Obviously this latter stage is only reached after a sufficient number of chemical facts have been ascertained. Thus we find that physical chemistry is a relatively youthful science, dating back, in fact, not much more than a single lifetime.

In the science of chemistry we are now accustomed to say that we *understand* a particular phenomenon, reaction, or property of matter when we can express it in terms of mechanics—*i.e.*, in terms of the fundamental concepts, time, length, and mass, and the derived concepts, velocity, acceleration, force, and energy. Our mental processes of analysis cannot go further back than that. When we state a chemical fact in such terms it must be clearly understood that we are not merely describing the fact, but, what is a great deal more important, we are showing that the fact or phenomenon itself is to be anticipated on the basis of a logical series of theorems or conclusions, for the truth of which we have already independent evidence, coupled very frequently with some new concept or generalisation which we had not hitherto recognised as existing, to which we give the name *assumption*. Science progresses by the cautious use of assumptions, which we must be careful to verify as rigidly as possible by appealing to experiment directly or indirectly.

Physical chemistry consists, therefore, of a body of theory or interrelated generalisations and principles, each of which has been tested in a variety of ways. The physical chemist applies such principles to chemical phenomena in his attempt to elucidate and understand these phenomena. The attitude of mind is well described by Sir J. J. Thomson when speaking of the theories of physics : " From the point of view of the physicist a theory of matter is a policy rather than a creed ; its object is to connect or co-ordinate apparently diverse phenomena, and above all to suggest, stimulate, and direct experiment. It ought to furnish a compass which, if followed, will lead the observer further and further into previously unexplored regions. Whether these regions will be barren or fertile, experience alone will decide ; but, at any rate, one who is guided in this way will travel onward in a definite direction, and will not wander aimlessly to and fro."

Matter and Energy.—From the point of view of the physical chemist there are two great concepts of which he makes constant use. These are *matter* and *energy*. All chemical changes—and all physical changes, for that matter—are concerned with the changes which matter can undergo. In very simple changes, physical changes, such as a change of state—*i.e.*, the vaporisation of a liquid or the melting of a solid—we are not dealing with the destruction or break-up of molecules into simpler units, but simply with a change in the spatial distribution of the molecules. When, however, we are dealing with the actual alteration of a molecule, the change is of a more fundamental character, and is spoken of as chemical. The two kinds of change often occur at the same time. For the sake of clearness, however, we keep them apart.

Parenthetically it is necessary to indicate the relationships which subsist between the three chemical "subjects"—inorganic, organic, and physical chemistry. From the standpoint of the physical chemist it is immaterial whether a given chemical change is classed as inorganic or organic. The physical chemist is equally interested in both. It will be readily seen, therefore, how inexact it is to say that there are three *kinds* of chemistry. As far as classi-fication of any reaction or process is concerned, it belongs necessarily either to inorganic or to organic chemistry. Physical chemistry, properly speaking, is not a third branch ; it is a new way of looking at and attacking the problems which inorganic and organic chemistry present.

We are speaking of material changes, and have been laying stress on the concept of matter. Equally important, though more subtle and less obvious, is the energy change which accompanies every change of matter. The significance of these concomitant energy changes appears when we turn to the question of the origin and cause of any chemical phenomenon or process.

Matter by itself is quite inert. But we must remember that we never deal with matter alone. We always encounter matter which contains a certain amount of energy, this amount depending upon the particular kind of matter and the external conditions, such as temperature and pressure. The amount of energy in a molecule—its energy content, as it is called—may vary very much from that of neighbouring molecules even in the same piece of material. It is natural, therefore, to seek in this energy content the source and origin of the chemical change which the matter undergoes. It is only the molecules which are surcharged with energy, as it were, which exhibit chemical reactivity—*i.e.*, possess the capacity of entering into chemical reactions. Energy exchanges between molecule and molecule or between the surroundings and the molecules are, therefore, not simply the accompaniment of chemical change, but are the immediate cause of the change itself. That is why many substances have to be heated before they will react or change chemically; as the temperature is raised, the energy content of some of the molecules likewise rises. It follows that we cannot be said to understand a chemical process until we know quantitatively the energy changes which have taken place at the same time. This is in general by no means an easy matter, and in fact there are no reactions about which our knowledge is complete in this sense, although we do know a good deal about a few reactions both in respect of matter and energy changes, and there are a very large number of reactions and processes about which we possess quite extensive knowledge from the point of view of matter alone.

The *rôle* which energy plays in chemistry is so fundamental that we must say a little more about it. From the standpoint of mechanics there are two *kinds* of energy, or, rather, energy manifests itself under two guises. These are called *kinetic* and *potential* energy respectively. By kinetic energy is meant the energy which a particle—*e.g.*, a molecule—possesses in virtue of its *motion*. By potential energy is meant the energy which a particle possesses in virtue of its *position* both with respect to other particles (molecules), and also with respect to its own component parts or atoms.* To understand this more clearly, it will be necessary to discuss in a little detail later on the question of the movement of molecules, their structure, and the forces of attraction between them.

Besides these two kinds of mechanical energy we have also to recognise different *sorts* or *types* of energy, which play a part in chemical and physical phenomena. These types of energy are heat energy, light energy, electrical energy, and magnetic energy. All these types

---

* A simple illustration of what we mean by kinetic and potential energy is afforded by a swinging pendulum. The pendulum tends to come to rest owing to the attraction of the earth upon it—*i.e.*, its stable position is the lowest position of its swing, for at the lowest position it is nearest the earth. The pendulum therefore possesses potential energy with respect to the earth. There is always an attractive force in operation when we find potential energy. At its highest point—*i.e.*, at the end of the swing—the pendulum is further from the earth than at the position of final rest; its potential energy is greatest at the end of the swing and least when the bob is passing through the lowest position. At the same time the pendulum is in motion and consequently it possesses kinetic energy. It is obviously moving with its greatest speed through the lowest position, and consequently just at this point its kinetic energy is a maximum. At the end of the swing the bob ceases to move for a moment, and there its kinetic energy is zero. So that in this case the potential energy is greatest where the kinetic energy is least, and *vice versâ*.

of energy must ultimately be expressible in terms of mechanical energy—*i.e.*, as kinetic energy, or potential energy, or both. With certain restrictions, all types of energy are interconvertible. The science which deals with such interconversions is called thermodynamics. This science represents a very generalised kind of thinking—as opposed to the " particularised " ideas involved in the concept of molecules and atoms—and it will be impossible for us to do justice to the very great importance and usefulness of thermodynamics for chemistry within the scope of this volume. In the concluding chapter, however, we shall make use of one thermodynamical principle—the well-known principle of conservation of energy, in connection with the heat evolved or absorbed in a chemical change. The greater part of what we propose to consider will consist of a survey of physico-chemical problems, primarily from the standpoint of the material changes involved, and only secondarily from the standpoint of the energy changes, not because these are less important, but because they are more subtle.

APPENDIX TO CHAPTER I.

SOME DEFINITIONS IN MECHANICS.

As we have been laying some stress upon the importance of attacking a chemical problem from the point of view of mechanics, it may be well to give a brief summary of the various mechanical terms and quantities which it is necessary to employ.

As already stated, the three fundamental mechanical concepts are *mass, length,* and *time.* It is necessary, of course, if we wish to express these concepts quantitatively, to possess some system of units. It is usual to express mass in terms of *grams,* length in terms of *centimetres* (cms.), and time in terms of *seconds.* This system of units is known as the centimetre-gram-second system, or, more briefly, the *c-g-s.* system. The corresponding algebraic symbols are m for mass, l for length, and t for time. Thus if we say that a body has a mass m, we mean that its mass is m grams. If the distance travelled by a body be l in time t, we mean that it has travelled l cms. in t seconds. Naturally we can express derived quantities in terms of the same units, though in some cases it is convenient, for the sake of brevity, to attach a particular name to the derived quantity and to the unit by which it is measured.

The first derived concept in mechanics—that is, an idea derived from and based on two or more of the fundamental concepts, mass, length, or time—is *velocity,* the velocity of a body or particle being defined as the distance or length travelled by the body or particle in unit time—*i.e.,* in one second. The algebraic symbol for velocity is u. Expressed in the *c-g-s.* system, we would say that a body travelling with a uniform velocity u traverses u cms. in one second. We say, for example—using another system of units which, however, are not employed in chemical problems—that a train has a velocity of sixty miles an hour, and we mean by that that it covers the distance sixty miles in the time one hour. We are thus able to express *quantitatively* the concept of velocity by making use of the two fundamental concepts length and time, these being expressed in certain units.

Again, in the above case we are evidently thinking of uniform or steady velocity. Let us now consider a body in motion with a velocity which is changing with the time. The velocity may be either increasing or decreasing with the time. In such a case we say that the body is suffering an *accelera-tion*. The acceleration is regarded as positive when the velocity of the body increases with the time, the acceleration being regarded as negative when the velocity of the body decreases with the time. Acceleration is, therefore, rate of change of velocity with time. Acceleration is, therefore, a less funda-mental quantity or concept than velocity, for we can define acceleration in terms of velocity and time, and we have seen already that velocity itself can be defined in terms of length and time, so that evidently acceleration can ultimately be defined in terms of length and time. We say that a body has an accel-eration *a* when its velocity is decreasing or increasing by *a* cms. per second per second. A train starting from rest has a positive acceleration, say, of so many cms. per second. After a time its velocity becomes uniform, and it no longer possesses any acceleration. As it approaches the next stop, however, its velocity diminishes—*i.e.*, it possesses a negative accelera-tion until its velocity falls to zero at the platform. Thus we can build up a relatively involved idea, such as that of acceleration, provided we have got some mental grasp of the two concepts length and time. The most important case of acceleration met with in physics is that known as the acceleration of gravity, which will be referred to later. The usual symbol to express the acceleration of gravity is *g*.

We can now go a step further and introduce a new derived concept—that known as *force*. Force is usually denoted algebraically by the symbol *F*. On the *c-g-s*. system of units, the unit of force is called the *dyne*. That is, we can speak of a force of so many dynes. Sometimes also we use the term gram to denote force, though it is a little confusing at first sight, since the gram is the unit of mass. As a matter of fact, a force of one gram is identical with 981 dynes. The use of the term gram in connection with force will be discussed later. For the present we shall keep to the dyne as the unit of force. One might be inclined to think that force was about as fundamental a thing as mass or length of time, and, indeed, at first sight force might not seem to be connected at all with these latter concepts. When, however, we get away from our in general very vague idea of force as something that can do something, we find that force can be defined in terms of mass and acceleration, being simply the product of these two quantities. Acceleration, we have seen, is definable in terms of length and time, and hence force is definable in terms of mass, length, and time. To get at this idea, we might imagine the case of a stone falling without friction under the attraction of the earth's gravitation. Gravitation is a force in virtue of which all bodies are attracted or tend to be drawn towards one another. We usually speak of gravitation in connection with the force of attraction or pull of the earth upon all bodies upon it or in space. When the stone, suspended at some distance from the ground, is in its initial position of rest its velocity is obviously zero. On being released it falls towards the earth, and, what is particularly important to observe, it falls faster and faster as it approaches the ground—*i.e.*, it suffers positive acceleration, the so-called acceleration of gravity, which, as already mentioned, is denoted by the symbol g. This acceleration has been measured, and is found to be 981 cms. per second per second. This acceleration is not to be confused with the force, the force of gravitation, which pulls the stone down. The force, in fact, depends not only upon the acceleration experienced by the stone, but also upon the mass of the stone, being, as we have said, the product of the mass into the acceleration. A large stone has a bigger pull upon it—*i.e.*, there is greater force exerted upon it by the earth than is exerted upon a small stone, because the mass of the bigger stone is greater than that of the smaller one. It can be shown, however, by direct experiment that both stones fall at the same rate,—that is, with the same acceleration, namely, g. One may easily see that the larger mass has the larger force of gravitation acting upon it by simply making use of a balance or scales. If the large stone be placed in one scale pan and the small one in the other, it will be seen at once that the large one is drawn more strongly towards the earth, for the scale pan containing the large stone is drawn downwards. The principle of the balance depends, indeed, directly upon the existence of the force of gravitation. A balance, therefore, measures relative forces, not masses directly. It follows from the experiment with the balance that force depends upon mass and upon acceleration ; in fact, upon the product of the two. If we denote mass by m and acceleration by a, then we can write the equation : $F = ma$.

In the particular case in which we are dealing with the acceleration of gravity which is denoted by g, we can write : $F = mg$, where F stands here for the force of gravitation. If we have two bodies whose masses are m and m_1 respectively, it follows that the gravitational force acting upon each of them— namely, F and F_1—will be given by the two expressions—

$$F = mg$$
$$\text{and } F_1 = m_1 g \;;$$
$$\text{or } \frac{F}{F_1} = \frac{m}{m_1},$$

which is simply an algebraic way of saying that the gravitational forces are directly proportional to the respective masses of the bodies compared.

Now we are very accustomed to use the term *weight* in connection with operations with a balance or scales. We must distinguish carefully between mass and weight. It must be clearly borne in mind that when we speak of weight—a body having such and such a weight—we are really dealing with and measuring force—*viz.*, gravitational force—and a pair of scales is essentially an instrument for comparing forces of gravitation. The term weight is identical with gravitational force, so that if we denote weight by the symbol W, it follows that we can write W and W_1 in place of F and F_1. By substituting these symbols in the equation obtained above we get—

$$\frac{W}{W_1} = \frac{m}{m_1},$$

which is the same thing as saying that the weights of two substances are in proportion to their masses. When, therefore, we say that the weight of a body is one gram, we mean that its mass is one gram. It is unfortunate that the word gram was ever used in connection with weight at all. In what follows we shall always denote mass by grams and force by dynes.

To indicate still further the distinction between weight (gravitational force of attraction exerted upon a body) and the true mass of a body, let us imagine ourselves carrying out the operation of weighing with a balance at the surface of the moon instead of upon the surface of the earth. Since the moon is a smaller body than the earth, its force of attraction or gravitation upon any piece of matter upon its surface is very much less than the force of attraction upon the same piece of matter when it is upon the earth's surface. The piece of matter is the same in both cases—that is, it has the same mass, but its weight —i.e., the pull or force exerted upon it—is very much less when the piece of matter is upon the moon's surface than when it is upon the earth's surface —i.e., the weight of a body may vary, but its mass cannot.

We have been speaking of force—force of attraction between bodies. There is a further instance of this force which is of very great interest especially as it brings us to a new concept—viz., the concept of pressure. We say that the atmosphere exerts a pressure of 15 lbs. to the square inch on the earth's surface. We mean by that that the gravitational attraction of the earth upon the gaseous constituents of the atmosphere exerts a pull—i.e., a force—of 15 lbs. weight over every square inch of the earth's surface. Pressure is therefore not a new concept, being simply a force acting over a unit area ; or, more briefly, pressure is force per unit area. On the *c-g-s.* system the unit of pressure, which is called a *bar*, is one dyne acting over or upon one square centimetre. Pressure can, therefore, be referred to the three fundamental concepts mass, length, and time, for force itself, we saw, was thus reducible, and area is simply the product of two lengths.

In the above case we have come upon the idea of pressure based upon a force of attraction exerted by the earth. As a matter of fact, for the purposes of physical chemistry, there is a far more important case to be considered, in which pressure or force per unit area manifests itself when one body in motion collides with another body which may either be in motion or at rest. In this case we are not thinking of the gravitational pull or force of attraction at all. Naturally, as long as we experiment upon the earth's surface, we cannot eliminate gravitation, but in very many cases the effect produced by gravitation is negligibly small compared with the effects produced in other ways. This is true when we are dealing with a number of rapidly moving bodies or particles which are colliding with one another. The effect we are talking about now is to be clearly distinguished from the " static " pull of the earth's gravitation upon a mass of gas—say, the gaseous atmosphere. The effect we are considering just now is a " dynamic " one—*i.e.*, it is the result of motion. In the act of collision the bodies are said to exert a force upon one another. This often manifests itself by one or both of the two colliding bodies being driven in a new direction. The act of collision is known as *impact*. Although the force considered here as a result of collision is quite different in origin from that which we have been considering previously in connection with gravitational attraction, nevertheless the force of impact is still measurable in the same units—*viz.*, dynes— being, in fact, always reducible to the product of mass into acceleration. The force exerted by the impact of two bodies—which force when it is measured over unit area we call pressure—has its origin and magnitude in the velocity and direction of the colliding bodies. This kind of force is of very great importance in connection with the behaviour of gases, as we shall see in the next chapter.

The force exerted upon one another by the impact of two moving bodies is very closely related to another mechanical concept—viz., *momentum*. When a body is moving we say that it possesses momentum. The momentum, which is usually denoted by the symbol M, depends on the mass of the moving body; the greater the mass, the greater the momentum. The momentum likewise depends on the velocity with which the body moves. The faster it is moving the greater is its momentum. The momentum of a body is, in fact, defined as the product of the mass by the velocity. Now what is the connection between the force exerted at impact and the momentum of the colliding bodies? We saw that force could be defined as the product of the mass into the acceleration of a freely moving body, and acceleration in turn is the rate of change of velocity per second. Hence force may be defined as the rate of change of momentum of the body—that is to say, the change of momentum per second. The force we are considering here is the force exerted by collision. To understand a little more clearly the connection between this force and the momentum which produces it, let us think of a cannon ball travelling at a steady rate towards a target. We are supposing that the air friction is zero, and also that any gravitational effects are negligibly small. If the velocity of the ball is u cms. per second, and the mass of the ball is m grams, the momentum possessed by the ball is mu. That is, $M=mu$.

Now suppose the ball strikes the target and rebounds—the ball and target being supposed perfectly elastic. On striking, the velocity of the ball is reduced to zero, the momentum being also necessarily reduced to zero. The ball has, therefore, suffered a negative acceleration. The ball now starts to rebound, its initial velocity being zero, and it again rises to its velocity u under the rather imaginary conditions we are assuming—*viz.*, that the elasticity of the ball and target is perfect. It has now a velocity u once more, but in the opposite direction. In the total process, therefore, the ball has suffered a change of momentum from the value mu to zero, and then up again to mu, the ball now travelling in the reverse direction. The total *change* in the momentum is, therefore, $mu + mu$, or $2mu$. Now, in the operation of striking the target, we know that force was exerted upon the target, and we come, therefore, to the conclusion that force in such a case is simply the rate of change of momentum of the moving body. We have even measured the force, for we can write the force F exerted on impact as equal to $2mu$. This expression will be found to be of very great service later. At first sight it might not appear to be in agreement with our first definition of force as mass multiplied by acceleration. The term u, however, really contains the acceleration. Thus, as already pointed out, in the act of striking the target the velocity of the ball falls from u to zero. On rebounding, the velocity rises from zero to u once more, but in the reverse direction. Acceleration is always change in velocity, so that the change in velocity is u on striking, and also u on rebounding, or $2u$ in all. That is, the force, which is mass multiplied by acceleration, is m multiplied by $2u$, or the force is $2mu$ as before.

Now that we have got some idea of force, let us pass on to find out its bearing upon another mechanical magnitude—viz., *work*.

The term work as ordinarily employed is used very vaguely ; but in the sense in which it is employed in mechanics its significance is quite sharply defined. When a force is exerted—say, by one's hand—upon an object, and that object is thereby transported without friction along a certain distance, then the work done by the hand is defined as the product of the force applied into the distance throughout which the force was kept on. Mechanical work is made up, therefore, of two factors—force and length. Now force, as we have seen, is definable in terms of mass, length, and time, and hence work also is definable in terms of these three fundamentals. We can also look at the work done by a body or agent of any kind from a slightly different point of view—*viz.*, from the point of view of the *energy* expended in doing the work. Work cannot be done without the expenditure of energy. When an isolated body does work it necessarily loses energy, and the energy lost is exactly equivalent to the work done. Since work done could be used to measure energy lost in such a case—that is, when all the energy is converted into work and not into something else, *i.e.*, some other form of energy—it follows that whatever terms or units we may employ to express work, the same units will express energy. Hence energy—*i.e.*, gain or loss of energy—can be expressed ultimately in terms of mass, length, and time. The unit amount of work will obviously be done when a unit force acts on some body, and moves it without friction through a distance of one centimetre. That is, the *c-g-s.* unit of work is the dyne-cm., or, to give it its more usual name, the *erg*. It follows from what has been said that mechanical energy may be expressed in ergs. This same unit is employed whether the energy be kinetic or potential—terms which we have already discussed in the text. What is particularly important is the distinction between work or energy and force. When a stone of mass m grams falls ten cms. towards the earth the force acting upon it is the *force* of gravitation, which we may denote by the symbol G; the *work* done upon the stone by the earth's attraction is 10 G, which will be expressed in ergs, provided G itself is expressed in dynes. From what has been said in the text about potential energy, it will be readily understood that in falling 10 cms. towards the earth the potential energy of the stone is diminished by the quantity 10 G.

We have already seen that the force of attraction is simply the mass of the body multiplied by its acceleration. If the force of gravitation is considered, and the mass of the body which falls through 10 cms. is, say, 100 grams, then we have the relation, $G=mg$, where g is the acceleration of gravity—g has been shown by experiment to be 981 cms. per second per second. Hence the force acting on the mass of 100 grams is 100×981, or 98,100 dynes. If under this force the body falls 10 cms., the work done by gravitation upon the body is $98,100 \times 10$, or 981,000 ergs. This number of ergs likewise expresses the decrease in potential energy of the body with respect to the earth's centre.

In connection with the other kind of mechanical energy—*viz.*, kinetic energy—it will be recalled that a body only possesses this if in motion. For reasons which need not be gone into, it has been found that the kinetic energy of a moving body may be expressed by the product of one-half of the mass of the body into the square of its velocity. If we denote the mass of a body by m and its velocity at the instant considered by u, then its kinetic energy is $\frac{1}{2}mu^2$. If the mass is expressed in grams and the velocity in cms. per second, then the kinetic energy is expressed in ergs. Thus a body of 100 grams in mass moving with a velocity of 1,000 cms. per second possesses kinetic energy amounting to $\frac{100}{2} \times (1,000)^2$, or 50,000,000 ergs. The kinetic energy of a body must be clearly distinguished from its momentum, which, we saw, was the product of its mass into its velocity. Thus in the case of the body of 100 grams mass moving with a velocity of 1,000 cms. per second, its momentum is $(100 \times 1,000)$, or 100,000 gram-cms. per second.

From the foregoing extremely sketchy outline it will be clear that terms such as work, energy, pressure, force, momentum, acceleration, and velocity, are all closely linked together, being all derivable from the three fundamental concepts, mass, length, and time.

Now the magnitudes which we have been considering may be all experimentally examined with materials which are of sufficient size to be easily perceived by our senses without the aid of any special apparatus. Cannon balls, targets, trains, and stones are all things large enough to be seen by the naked eye. The science of mechanics deals in the first place with such, but it is not limited merely by size ; and by making use of the same ideas and principles it has been found possible to advance a considerable way into the problem of the mechanism of what is taking place in systems built upon so minute a scale as not to be directly observable even by the most delicate microscope. In other words, the principles of mechanics can be applied to the behaviour of molecules, and with suitable precautions to the very structure of molecules. In short, the principles of mechanics have been applied, and applied with a very considerable measure of success, to the elucidation of problems connected with the ultimate constitution of matter and the changes which matter can undergo. It is for this reason that it was necessary to consider mechanical principles at all before passing on to the problems of physical chemistry. It may, perhaps, be mentioned that when we make such applications of mechanics to the behaviour and properties of molecules, we are employing a series of considerations which are grouped under the general title, " the kinetic theory of matter."

CHAPTER II.

IT is well known that matter is capable of existing in three physical states—the solid, the liquid, and the gaseous ; and, further, that transition from one state to another is possible under certain conditions. In general, it is true to say that at low temperatures substances are found in the solid or liquid form, whilst at higher temperatures they become gaseous.

The changes from one state to another— *i.e.*, the processes of *melting, solidification, vaporisation,* and *condensation*—are all regarded as physical, because no fundamental or chemical change occurs in the nature and structure of the molecules concerned as a result of such processes. The real distinction between the three states is one of spatial distribution of the molecules—that is to say, the arrangement, distance apart of the molecules, and their type of motion, differs, say, in the solid state very considerably from the arrangement, mode of motion, and distribution of the molecules in the gaseous state ; but at the same time the substance is chemically unchanged. The substance the molecule of which is represented by the formula H_2O exists in the liquid form as water, in the solid form as ice, and in the gaseous form as water vapour or steam.

Although spatial distribution is the fundamental distinction between the three states of matter, it must not be imagined that in all cases chemical change, of a relatively feeble kind, is wholly absent. In the gaseous state, cases are known in which at high temperatures the actual molecules are broken up or dissociated into atoms. We are not thinking of this kind of chemical change, however, but of the more feeble kind known as *association* or *polymerisation* of molecules themselves, which is known to occur in the case of some substances, especially in the liquid state.

Thus in the case of water it is now fairly certain that the solid form, ice, consists mainly of molecules having the formula $(H_2O)_3$—*i.e.*, each molecular unit in ice consists of three simple H_2O molecules united together. It is equally certain that liquid water at ordinary temperature consists mainly of molecules having the composition $(H_2O)_2$—*i.e.*, individuals produced by the cohesion of two simple molecules. In liquid water there are also a very small number of ice molecules, $(H_2O)_3$, and a considerably larger number of the simple molecules, H_2O. As the temperature of liquid water is raised, say up to near the boiling-point, the ice type of molecule, $(H_2O)_3$, has entirely vanished, the $(H_2O)_2$ type has also greatly diminished, and there is a corresponding very large increase in the percentage of simple H_2O molecules. The changes considered here are certainly chemical changes, for they can be represented by ordinary chemical equations, *viz.*:—

$$(H_2O)_3 = 3H_2O.$$
$$(H_2O)_2 = 2H_2O.$$

The gaseous state of the same substance—*viz.*, water vapour or steam—consists almost entirely of the simplest molecules, H_2O. As a matter of fact, water is a particularly complex substance as far as molecular individuals are concerned. Many substances—such as benzene, for example—exist in the simplest form of molecules, C_6H_6, even in the liquid state, these molecules being identical chemically with those of the vapour of benzene. In all cases the gaseous state represents a simpler state of molecular affairs than is found in the liquid or solid state. It is obvious, therefore, that the gaseous state affords the most promising field for consideration, because the simplest molecular conditions are met with there. We shall therefore devote some further consideration to the gaseous state in the present chapter, for, as a matter of fact, our knowledge of the gaseous state is very much more complete than is our knowledge of the liquid or solid state.

It is very necessary for the reader to appreciate the general nature and scope of the problem which is presented to us by the existence of the three states of matter. It is not sufficient simply to know the bare fact, as a result of observation, that such states do exist ; the aim of molecular physics has long been to understand and explain on *mechanical grounds* the molecular mechanism which underlies these modes of molecular distribution, and the conditions which determine the transition from one state to another. Of course nothing like finality has been or can be attained in such investigations. It is of interest, however, to review briefly some of the conclusions which have been arrived at, and for this purpose we shall begin with some phenomena characteristic of the gaseous state itself.

The Gaseous State.—The simplest and most important relationship or law which has been found to hold good for a substance in the gaseous state is that known as *Boyle's Law.* This law was discovered experimentally by Boyle in 1660, and concerns itself with the relation which exists between the volume of a given mass of gas and the pressure it exerts upon the walls of the containing vessel. Boyle's law states that *the volume occupied by a gas is inversely proportional to the pressure exerted by it,* or, what is the same thing, the pressure exerted upon it. Another way of stating the law is to say that the product of the volume into the pressure is a constant. By the term " constant " we mean a fixed numerical value which retains its value, no matter how much the pressure, and consequently the volume, of the gas may be altered. It is assumed that the temperature of the gas is unchanged during the experiment.

Let us suppose that we have a certain mass of gas, enclosed in a cylinder which is fitted with a gas-tight but easily movable piston upon which we can place weights, thereby altering the pressure on the gas. Let us suppose that we start with the gas thus enclosed (Fig. 1). The pressure on the piston, before any weights have been added, is simply one atmosphere, the earth's atmosphere, which is equivalent to a weight of 15 lbs. on every square inch of the piston surface. If the piston is stationary—*i.e.*, is in a state of equilibrium—it is obvious that the pressure ex-erted by the gas itself upon the lower face of the piston is just one atmosphere. Let us suppose that in this initial state the volume of the gas is 1,000 ccs. —*i.e.*, one litre. Now

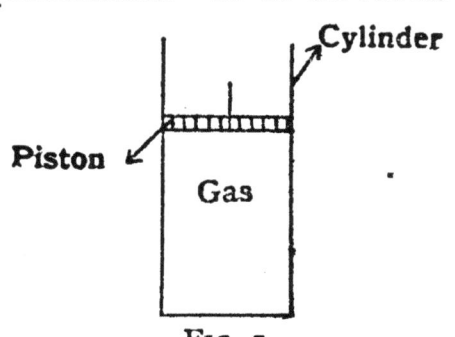

FIG. 1.

let us put a number of weights upon the piston, so as to be equivalent to an additional pressure of 15 lbs. to the square inch. (If the area of the piston head had been one square inch, then we would simply have had to put on a 15-lb. weight to increase the pressure by 15 lbs. per square inch. If the area of the piston head were 10 square inches, we would have to put on 150 lbs. in all to produce the same increase in pressure.) Now when this extra pressure of 15 lbs. per square inch has been put on, making a total pressure of 30 lbs., or two atmos-

pheres, it will be found that the piston has
descended, the gas being compressed and occupying a smaller volume than before, and, indeed, the piston will again come to rest when
the volume of the gas has diminished by one-
half—*i.e.*, when its volume is 500 ccs. On
adding a further 15 lbs. pressure, making in all
a pressure of 45 lbs., it will be found that the
gas now occupies a volume of 333.3 ccs. On
adding a further 15 lbs., making in all 60 lbs.
pressure, the volume of the gas will be found
to be 250 ccs. It is obvious from these figures
that the volume is inversely proportional to
the pressure exerted by or upon the gas, as the
following table shows. [*Note.*—When we say
that a quantity x is inversely proportional to
another quantity y, we mean that $x = \dfrac{k}{y}$
where k is a proportionality factor or constant.
This equation may also be written, $xy = k$—
i.e., the product of x and y is a constant,
howsoever x and y may vary.] We denote
pressure by P and volume by V.

VOLUME AND PRESSURE OF A GAS. BOYLE'S LAW.

Pressure on Piston (P).	Volume of Gas (V).	P × V.
15 lbs. (the earth's atmosphere)	1,000 ccs.	15,000
30 ,, ,,	500 ,,	15,000
45 ,, ,,	333.3 ,,	15,000
60 ,, ,,	250 ,,	15,000

Boyle's law is therefore often stated in the form—

$$PV = \text{a constant} = k.$$

We have been thinking of a " certain mass " of gas, which mass occupied 1,000 ccs. at the temperature in question, and at one atmosphere pressure. If we had used twice as much gas—*i.e.*, twice the mass—the volume of it would have been 2,000 ccs. at one atmosphere pressure, and therefore the constant k would have had the value 30,000 throughout. It is obvious, therefore, that the numerical value of k depends upon the *mass* of gas chosen for the experiment. Further, the value of k depends upon the temperature; for, as we shall point out in the next paragraph, the volume of a given mass of gas varies with the temperature, in the sense that the volume increases as the temperature rises, and diminishes as the temperature falls.

The Law of Gay-Lussac or Charles.—This is the second fundamental law of gases. According to this law, *the volume of a given mass of gas kept at constant pressure increases by a definite fraction of itself for every degree rise in the temperature of the gas.* The fraction here referred to is usually denoted by the symbol a. Consider once more the cylinder of gas (Fig. 1). The volume is 1,000 ccs. under the pressure of one atmosphere at a given temperature. Now let us warm up the gas 10 degrees. The gas expands, the piston being pushed outward until a new position of equilibrium is reached cor-

responding to the higher temperature. Notice
that the pressure in this experiment is kept
constant — *viz.*, one atmosphere — for no
weights are added, and the piston is free to
move. That is, the pressure exerted by the
gas is one atmosphere, a constant value
throughout the experiment. It will now be
found that the volume occupied by the gas is
1,036.6 ccs. There does not seem to be, at
first sight, any relation between this number and the volume—*viz.*, 1,000 ccs.—which
the gas occupied at the initial temperature,
10 degrees lower than the final one. Nevertheless there is a very simple relation between
them which Gay-Lussac and Charles discovered. The law which has already been
stated in words may also be expressed algebraically in the following way—

$$V_t = V_o(1 + a\,t).$$

Where t denotes the increase in temperature
which the gas experiences at constant pressure, V_o is the initial volume of the gas at the
lower temperature, V_t is the final volume of
the gas at the higher temperature, and a is a
coefficient or factor which has the same value
for all gases. We can calculate the value of
a from the above experiment. Thus in the
above case $V_o = 1,000$ ccs., $V_t = 1,036.6$ ccs.,
and $t = 10$, the *increase* in temperature. On
substituting these values in the above equation, we find that $a = 0.00366$, or $a = \dfrac{1}{273}$.
The law of Gay-Lussac or Charles—which ex-

presses an experimental *fact*—means that if we start with a gas occupying volume V_o, and heat it up one degree—*i.e.*, $t=1$—then its new volume V_t will be given by the expression—

$$V_t=V_o(1+a)=V_o+aV_o,$$
$$\text{or } V_t-V_o=aV_o.$$

But V_t-V_o is the increase in the observed volume due to the rise in temperature of one degree at constant pressure. So that we can say—

observed increase in volume per degree $=aV_o$.

Now if we divide the observed increase in volume by V_o, the original volume, we get a quantity which means the increase in volume expressed as a *fraction* of the original volume ; or, more briefly, we get the fractional increase in volume per degree. Obviously these words can be expressed algebraically thus—

$$\frac{\text{observed increase in volume per degree}}{V_o}=a,$$

so that a stands for the fractional increase in volume per degree rise in temperature. The numerical value of a, we have seen, is $\frac{1}{273}$. Suppose, therefore, that we start with 273 ccs. of gas at a given pressure and at zero centigrade, the freezing point of water, and lower the temperature by one degree, then the volume will decrease by $\frac{1}{273}$rd part of itself— that is, by 1 cc. Hence the volume occupied by the gas at one degree below zero centigrade will be 272 ccs. Again, lower the temperature by one degree. The volume will now be 271 ccs. Obviously, if we lower the temperature 273 degrees below the freezing-point of water, the gas would not occupy any volume at all. Although this, strictly speaking, is an impossible result, nevertheless it indicates that there is something unique about the exceedingly low temperature which we have just been considering—*viz.*, 273 degrees below zero centigrade. As a matter of fact, we have arrived, on the basis of the Gay-Lussac or Charles law, at the idea of a *lowest possible temperature*, known as the *absolute zero*, or starting-point on the so-called absolute scale of temperature measurement. The starting-point on the absolute scale is therefore 273 degrees below the freezing-point of water, so that, expressed on the absolute scale of temperature, we would say that water freezes at 273 degrees. The absolute scale of temperature, or, as it is more usually called, the absolute temperature, is denoted by the symbol T, to distinguish it from the centigrade temperatures which are denoted by *t*. It is obvious that the connection between the two scales is given by the relation $T = t + 273$.

From what has been said, it follows that the volume of a gas at constant pressure is directly proportional to the absolute temperature of the gas—*i.e.*, $V = k_1 T$, where k_1 is a proportionality constant. This, in fact, is an alternative way of stating the law of Gay-Lussac or Charles. Now at constant temperature we know from Boyle's law that $PV = a$ constant.

But we have just seen that V at a constant pressure is directly proportional to T. On combining these two laws we obtain the relation $\frac{PV}{T}=$a constant, or PV=constant$\times$T.

This expression, which comprises in a brief form both experimental laws, is known by the general name, the gas law. It is not quite complete, however, as it stands, for we have not in any way defined the mass of the gas considered, and the numerical value of the proportionality factor or constant depends upon the mass chosen. It is usual to think always of one grammolecule as the mass of gas considered. The grammolecule is simply the molecular weight expressed in grams—*e.g.*, 32 for oxygen, 28 for nitrogen, 2 for hydrogen, and so on, taking the mass of the gram atom of hydrogen to be arbitrarily one gram. If we experiment with, say, 2 grams of hydrogen gas, and measure its volume, say, in litres at various temperatures on the absolute scale, and at various pressures (expressed in atmospheres), then it is found that the above expression takes complete account of the behaviour of the gas, the numerical value of the constant being 0.082. The units in which this is expressed are litre-atmospheres. This very important quantity is always denoted by the symbol R, and is known as *the gas constant*. Exactly the same value for R is obtained if we experiment with 28 grams of nitrogen gas, or with 32 grams of oxygen gas, and measure the volume—pressure—temperature relations. The value of R is therefore fixed—*i.e.*, it is independent of the chemical nature of the gas examined, provided only that we work with one grammolecule of the gas. The real meaning of this will be clearer when we have discussed the Avogadro hypothesis. To sum up, however, as far as we have gone, we can say that the behaviour of gases is taken account of by the simple expression known universally as *the gas law*—

$$PV = RT.$$

The Avogadro Hypothesis.—This is the third great concept in connection with the gaseous state. The hypothesis was put forward about a century ago by Avogadro, and its importance for chemistry can scarcely be overrated. The hypothesis may be stated thus : *Equal volumes of all gases at the same temperature and pressure contain the same number of molecules.* This relation is spoken of as an hypothesis rather than as a law, because it cannot be tested directly, say by actually counting the molecules. Nevertheless, its validity has been proved indirectly and most rigidly. It is essential to grasp the significance of this hypothesis. In the first place, equal volumes of different gases at the same temperature and pressure will certainly *not* mean equal masses of gases. Yet there are the same number of molecules present in each case. The molecules of different gases have different masses. Now the so-called grammolecules, or grammolecular weights, of different gases simply denote the relative masses of the different molecules, expressed in grams, the atomic mass of hydrogen being arbitrarily taken as unity. That is, the grammolecule is the same multiple of the actual and true mass of every molecule—the multiple being an enormous number, now known as the Avogadro number or constant. In other words, the grammolecular weights of different gases refer to the same number of actual molecules, and hence on the Avogadro hypothesis the grammolecular weights of different gases should occupy the same volume at a given temperature and pressure. As a matter of experimental fact they do so, the volume at zero centigrade and at one atmosphere pressure being 22.3 litres. It is from this observation that the idea of the gas constant R being independent of the nature of the gas is derived. If we always work with one grammolecule of any gas, then from Avogadro's hypothesis the value of the volume V will be the same for any gas at a given temperature and pressure. As already stated, the volume occupied by one grammolecule of any gas at $O^{\circ}C$ and one atmosphere pressure is 22.3 litres. The zero centigrade is 273 degrees absolute. We have, therefore, $V = 22.3$, $P = 1$, and $T = 273$. On substituting these values in the gas law— *viz.*, $PV = RT$—we obtain—

$$1 \times 22.3 = R \times 273 ;$$

whence $R = 0.082$ litre-atmospheres.

We have now considered the two fundamental experimental laws relating to gases— *viz.*, the law of Boyle and the law of Gay-Lussac or Charles; we have also considered the great generalisation known as the Avogadro hypothesis; and we have seen that all three are expressed or summed up in the single expression, $PV = RT$, where R refers to one grammolecule, and has the numerical value 0.082. We have now to see if these relationships or laws can be accounted for on the basis of mechanics as applied to molecular systems —the so-called kinetic theory of matter; for,

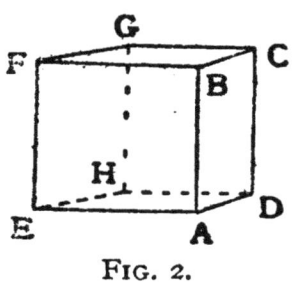

FIG. 2.

as pointed out in Chapter I., it is only when we are able to find a mechanical basis for a phenomenon or experimental law that we can be truly said to understand it.

We shall first of all consider the law of Boyle. To deduce it on the basis of the kinetic theory, we proceed as follows.

Consider a little box of unit dimensions— *i.e.*, one cm. cube, containing gas. The molecules of the gas are all supposed to possess the same velocity—*u* cms. per second. The actual mass of each molecule is *m* grams.

The molecules move in all directions in the actual case; but for the sake of simplicity we shall consider them as divided into three sets —*viz.*, one-third of them moving back and fro from the face ABCD to EFGH (Fig. 2), one-third passing between the faces ABEF and

GCDH, and one-third passing between the faces BCGF and ADHE. Let us consider one of these sets, say the set which strike and rebound from the walls or faces ABCD and EFGH. Let us further consider one molecule in this set. The molecule starts, say, from the face ABCD, and moves in a straight line to EFGH, which it strikes and from which it rebounds. The distance between each opposite face is one cm., and each time the molecule traverses one cm., therefore, it collides with a face, either EFGH or ABCD, upon which it exerts a force due to the impact. The velocity of the molecule is u—i.e., it traverses u cms. per second. Hence it collides with the opposite faces u times per second. That is, it collides with one of the faces, say the ABCD face, $\frac{u}{2}$ times per second. Now consider a single collision a little more closely. The total change in momentum when the molecule strikes and rebounds is, as we have seen in the Appendix to Chapter I., the quantity $2mu$. It has already been pointed out that force exerted at impact is the total change in momentum, so that $2mu$ is the force—outwardly directed—exerted by a single molecule in a single collision. Now a single molecule makes $\frac{u}{2}$ collisions per second on each face. Hence the total force exerted by a single molecule upon a face of the box is $\frac{2mu^2}{2}$, or simply mu^2. The *pressure* exerted by the gas on any of the faces is simply the total force exerted per unit area. If there are n molecules in all in the unit cube, then there are $\dfrac{n}{3}$ travelling between any pair of opposite faces. But the force exerted per second by a single molecule is mu^2. Hence the force exerted over the whole face, say ABCD, is mu^2 multiplied by the number of molecules which strike the face per second. That is, the total force exerted upon a single face is $\dfrac{nmu^2}{3}$. But the face is of unit area, and force per unit area is pressure. Hence the pressure exerted by the gas is given by the same expression—that is,

$$P = \frac{nmu^2}{3}.$$

Since we have measured m in grams and u in cms. per second, it follows that P is given in terms of dynes. Now, since m is the mass of a single molecule, and n is the number of molecules in unit volume, the unit cube, it follows that the quantity nm denotes the total mass of gas in the unit volume. But the mass of any substance which occupies unit volume is the *density* of the substance, and is usually denoted by ρ, so that we can write—

$$P = \frac{\rho u^2}{3}.$$

We have thus come to the conclusion that the pressure exerted by a gas is proportioned to its density and likewise to the square of the average velocity of the molecules. To arrive at Boyle's law—which, it will be recalled, holds good only so long as the temperature is maintained constant, during the volume and pressure changes—it is necessary to consider what we mean by *temperature* on the basis of the kinetic theory of matter.

According to the kinetic molecular theory, matter in the gaseous state consists of small particles or molecules which are moving rapidly about, colliding with one another and with the walls of the containing vessel. The number of collisions per second is enormously great. Further, we have reason for believing that the average distance of the molecules apart is great compared with the actual size of the molecules themselves. Of course there are many millions of molecules in one cc. of gas, and when we speak of their distance apart as large, we mean relatively to their own dimensions. We now come to a very important application of the idea of molecules in rapid and irregular motion. When heat is applied to a body—a gas, for example—the temperature of the gas is said to rise. Now temperature must be closely related to molecules and their motion. As a matter of fact, when heat —*i.e.*, heat energy—is absorbed by a body, part of the heat goes to increase the kinetic energy of the molecules. Now the kinetic energy of a molecule is $\frac{1}{2}mu^2$ where m is the mass of a molecule and u is its velocity. If the kinetic energy of the molecule is increased by warming, it means that the molecules of the warmed gas are flying about more rapidly than when the gas was cold. From the point of view of the kinetic molecular theory, therefore, *we find in the kinetic energy of the molecules a measure of the temperature of the substance.* We may see how this idea of the kinetic energy of the molecules can be used to measure the temperature of the substance, if we think of what happens on cooling a gas very greatly. As we go on cooling a body— *i.e.*, lowering its temperature—the motion of the molecules slows down, so that at a very low temperature indeed—a temperature that has not yet been realised experimentally, though we have come within a few degrees of it—the molecular motion will cease entirely. This unique state of affairs obviously represents a limiting condition, and, in fact, we regard it as identical with the absolute zero, or zero temperature on the absolute scale. This temperature is 273 degrees below the freezing-point of water. Since molecular motion, and therefore the kinetic energy of the molecules, increases or decreases as the absolute temperature of the substance rises or falls, it appears reasonable to regard the kinetic energy of an average molecule as a measure of the temperature, the absolute temperature, of the substance. It follows from this conclusion that if we keep a substance at a constant or fixed temperature, we are really keeping the average kinetic energy of each molecule likewise constant or fixed. Hence at a constant temperature we say that the quantity $\frac{1}{2}mu^2$ is constant. If this expression remains constant, it follows that u^2 itself, or the square of the average velocity of each molecule, is also constant.

Let us return now to the expression which we have obtained already for the pressure of a gas, namely—

$$P = \frac{\rho}{3} u^2.$$

If the temperature be kept constant, it follows that u^2 is constant, and of course the fraction $\frac{1}{3}$, being a pure number, is necessarily constant; so that we conclude that, *at constant temperature*, the pressure of a gas is directly proportional to the density of the gas. If now one gram of gas occupies a volume V c.cs., then $\frac{1}{V}$ is the density of the gas, and is, consequently, identical with ρ. Hence we can write the above conclusion in the form—

$$P \text{ is proportional to } \frac{1}{V}.$$

Instead of using the words " is proportional to," we can substitute the sign of equality and introduce a constant factor of proportionality, say k. The above statement can, therefore, be written thus—

$$P = \frac{k}{V},$$
$$\text{or } PV = k,$$

which is Boyle's law, deduced on the basis of mechanics as applied to a molecular system.

We now pass to a consideration of Gay-Lussac's or Charles's law. We have seen that this law may be stated thus: The volume of a gas maintained at constant pressure is directly proportional to the absolute temperature.

Let us write down again the expression already obtained for the pressure of a gas, *viz.*—

$$P = \frac{nmu^2}{3},$$

where *n* stands for the number of molecules per unit volume. If the actual volume of the gas is V c.cs., and there are N molecules present, it follows that there are $\frac{N}{V}$ molecules per unit volume. In other words, $n = \frac{N}{V}$. We can, therefore, rewrite the above expression in the form—

$$P = \frac{Nmu^2}{3V},$$

which may also be written thus—

$$P = \frac{2}{3V} \cdot \frac{Nmu^2}{2}$$

The quantity $\frac{mu^2}{2}$ is the kinetic energy of a single molecule. But the absolute temperature is proportional to the kinetic energy of each of the molecules. That is, the kinetic energy of the molecules is equal to k_1T, where k_1 is a constant or factor of proportionality, and T stands for the absolute temperature. We can substitute this for the expression $\frac{mu^2}{2}$ in the above equation, and obtain—

$$P = \frac{2}{3V} k_1 T . N,$$

$$\text{or } PV = \frac{2k_1}{3} T . N.$$

But $\frac{2k_1}{3}$ is a constant. Likewise, as long as we use the same mass of gas, N remains the same. We can write, therefore—

$$PV = \text{a constant} \times T.$$

If now P be maintained constant during an experiment in which T, and consequently V, are altered, it follows that we can write the above expression in the form—

a constant × V = a constant × T ;

or, V = a new constant × T ;

or, V is proportional to T,

which is the law of Gay-Lussac or Charles.

In the last place, we have to consider the Avogadro hypothesis. According to this hypothesis, equal volumes of different gases at the same temperature and pressure contain the same number of molecules.

To deduce this, let us consider two different gases in two separate vessels occupying the same volume and at the same temperature and pressure. The pressure exerted by the first gas may be written—

$$P = \frac{N_1 m_1 u_1^2}{3V}.$$

Where N_1 is the number of molecules of this gas in the volume V c.cs., m_1 is the mass of each molecule, and u_1 the average velocity of each molecule of this gas. Similarly, for the second gas we can write—

$$P = \frac{N_2 m_2 u_2^2}{3V},$$

where N_2 is the number of molecules of the second gas in the volume V, m_2 the mass of each molecule, and u_2 the average velocity of each molecule of the second gas. But the pressure exerted by each gas is the same. Hence—

$$\frac{N_1 m_1 u_1^2}{3V} = \frac{N_2 m_2 u_2^2}{3V},$$

or, $N_1 m_1 u_1^2 = N_2 m_2 u_2^2.$

Further, the temperature of both gases is the same—that is, the average kinetic energy of the molecules is the same. This is expressed by writing—

$$\frac{m_1 u_1^2}{2} = \frac{m_2 u_2^2}{2},$$

or, $m_1 u_1^2 = m_2 u_2^2.$

By combining the latter relation with the one obtained a few lines above—namely,

$$N_1 m_1 u_1^2 = N_2 m_2 u_2^2—$$

it follows that

$$N_1 = N_2.$$

That is, there are the *same* number of molecules in equal volumes of two gases at the same temperature and pressure. This is the Avogadro hypothesis.

None of the proofs given above are rigid. They serve to show, however, how we always attempt to understand and account for a physico-chemical phenomenon or relationship by finding its mechanical basis.

Compressed Gases and Liquids.—The gas law $PV = RT$ has been investigated in a very large number of cases with gases of different chemical nature, and under temperatures and pressures which exhibit a wide range. It has been found that, whilst this law is obeyed with very considerable exactness over certain ranges of temperature and pressure, nevertheless certain discrepancies manifest themselves. That is, neither Boyle's law nor the law of Gay-Lussac or Charles is really exact over the whole range of possible temperatures and pressures. The deviations are in general sensible when the gases are highly compressed, and especially at low temperatures. The kind of deviation from the law of Boyle, for example, actually found in practice, especially at low temperatures, is exhibited in the accompanying graph (Fig. 3). The horizontal line indicates the behaviour of a gas which rigidly obeyed Boyle's law, a so-called "perfect gas." The ordinate denotes the value of the product PV, the abscissæ representing the pressure P. According to Boyle's law, the product PV should be a constant at constant temperature—*i.e.*, the product should not vary with

increasing P. This is represented by the line
being horizontal. In actual cases, however,
we find that at low and up to moderate pressures the line representing PV is really horizontal ; but after the pressure has risen considerably the product becomes somewhat less
than it ought to be. This is shown by the
line falling below the level of the horizontal.
At still higher pressures the line turns upward,
and actually cuts the theoretical horizontal

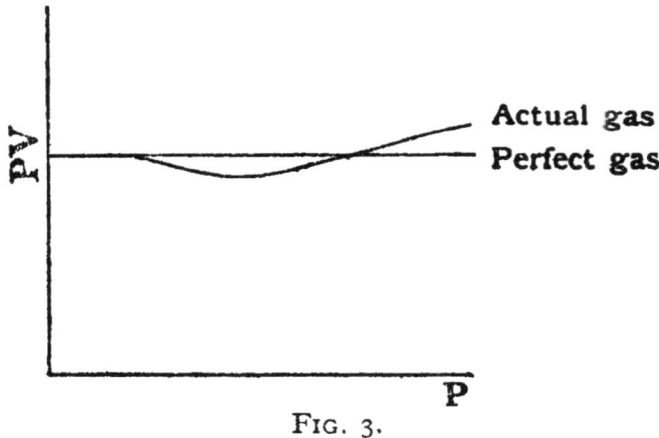

FIG. 3.

line, and now rises steadily as the pressure is
still further increased.

It was only natural that search should be
made for the cause of these deviations, and
eventually two disturbing factors or causes
were discovered. In the simple deduction of
Boyle's law it will be remembered that we
assumed that the size of the molecules themselves was negligibly small compared with the
total volume of the gas. The molecules, however, do possess a definite though small volume,

and molecules are practically incompressible. That is the first factor. Secondly, no account was taken of the forces of attraction between molecules, although when the molecules are close together these forces may become large. We can now explain the shape of the curve shown in Fig. 3. At low pressures the volume of the given mass of gas is necessarily large. That is, the force of attraction between the molecules is really negligible, and the volume occupied by the molecules themselves is also quite negligible compared with the total volume of the gas. Hence Boyle's law holds accurately, as shown by the curve being horizontal. As the pressure is increased, however, and consequently the volume diminished, the force of attraction begins to operate between the molecules, in virtue of which the molecules tend to be drawn together. That is, the gas contracts too easily, the attractive forces really aiding the external pressure. The result is that the volume diminishes rather more than would occur if no forces of attraction between the molecules existed. The net result is that the product PV becomes somewhat smaller than its theoretical value. That is, the line representing the value of PV falls below the horizontal. But as the pressure is still further increased, the molecules are now fairly close together, and the actual volume occupied by these little particles of matter is no longer negligible compared with the 'total volume. But the particles themselves cannot be compressed. That is, the gas now becomes not compressible enough. In other words, the volume does not diminish sufficiently on increasing the pressure by a certain amount. Hence the product PV now begins to increase, and eventually its value is much greater than that of the perfect gas. This is shown by the ascending part of the curve in the figure. These two disturbing factors—*viz.*, the size of the molecules and the forces of attraction between them—were first systematically examined by van der Waals, who attempted, with a very great measure of success, to modify the gas law—*i.e.*, Boyle's law—by the introduction of two so-called correction terms to account for these deviations of actual gases from the perfect gas law. Not only did van der Waals succeed in this, but at the same time he showed that his modified gas law applied fairly well to liquids, to which Boyle's law does not apply at all.

Briefly, the reasoning pursued by van der Waals is as follows.

He denoted the actual *in*compressible volume occupied by the molecules themselves by the symbol b. That is, b represents the volume which would be occupied by the gas if all the molecules were touching one another. The real compressible volume is therefore $(V - b)$, where V is the total observed volume of the gas. As a first step, therefore, to modify the gas law (*viz.*, $PV = RT$), van der Waals writes $P(V - b) = RT$. This, however, is not a sufficient modification. We have still to take account of the forces of attraction between the molecules. We have already pointed out that this force acts like an additional compressional pressure, which we may denote by the symbol π, so that the effective compressional agency is $(P + \pi)$, where P is the observed pressure of the gas as read on a manometer. The modified gas law is, therefore,

$$(P + \pi)(V - b) = RT.$$

This expression is undoubtedly a truer representation of the behaviour of a compressed gas or liquid than is the simple gas law itself. We have not made much progress, however, until we can evaluate π—*i.e.*, express it in terms of magnitudes which are measurable experimentally. Van de Waals writes π as proportional to the square of the density of the gas. That is,

$$\pi \propto \rho^2 = a\rho^2,$$

where a is a constant or proportionality factor which is characteristic of each gas, and is independent of the temperature and pressure of the gas. Now we have seen that the density can be written as the reciprocal of the volume, so that finally van der Waals' equation becomes.

$$\left(P + \frac{a}{V^2}\right)(V - b) = RT.$$

Van der Waals arrived at the conclusion that π, the cohesion tension or force between the molecules, reckoned per unit area, could be written as proportional to the square of the density somewhat in the following way. In the phenomenon of cohesion it is reasonable to assume that the cohesion will be greater the more closely packed are the molecules. That is, the cohesion π across any unit plane inside the gas is some function of the density of the gas. The cohesive force or force of molecular attraction will, in fact, depend upon the product of the number of molecules per unit volume on each side of the unit imaginary plane. Now density varies as the number of molecules per unit volume, so that the cohesive force will likewise depend upon the product of the densities of the gas on either side of the unit plane. But the gas is uniform in density, so that the densities on each side of the plane are identical. That is, the product of the densities is the same thing as the square of the density, or the square of the reciprocal of the specific volume.

We cannot go further into the question of the degree of the applicability of van der Waals' equation. Suffice it to say that it represents with remarkable fidelity the behaviour of compressed gases, and likewise of liquids. In particular, it affords a theoretical basis for the phenomenon first discovered by Andrews, and known now as the *continuity of the gaseous and liquid states*. Andrews found that there was a certain temperature, characteristic of each substance, above which it is impossible to liquefy the gas, no matter how great a pressure may be used. This temperature is called the *critical temperature* of the substance. For ether, the critical temperature is 193.8° centigrade ; for carbon dioxide it is 31.35° C. ; for water it is 360° C. The *smallest* pressure which will just bring about liquefaction at the critical temperature is called the *critical pressure.* For ether, the critical pressure is 35.6 atmospheres ; for carbon dioxide it is 72.9 atmospheres ; and for water it is 195.5 atmospheres. The volume of one gram of material at the critical temperature and pressure is called the *critical volume.* The reciprocal of this is called the *critical density.* The critical density of ether is 0.2625 ; for carbon it is 0.464; and for water it is 0.2078. When a substance is at its critical temperature and pressure, it is said to be *in the critical state.*

A few words further regarding the cohesion or force of attraction between molecules. It can be shown that the distance over which this force is effective is very small. That is, it is only in the close vicinity of a molecule that it can exert any attraction. In fact, the distance over which the attraction is measurable is only of the order 10^{-7} cm.—*i.e.,* one ten-millionth of a centimetre. It must not be thought, however, that because the distance over which the attraction is effective is small that the force of attraction itself and the effects produced by it are small. In liquids the molecules on the average are only separated by about five to ten hundred-millionths of one cm., so that they lie well within the range over which molecular attraction is effective. To give an idea of how great the cohesion is in liquids, a few values of π, the cohesion force reckoned per unit area in the liquid, are collected in the following table. The term π is often called the *internal pressure*. It must be remembered, however, that it is a tension rather than a pressure, drawing the molecules together.

Liquid.	Internal pressure at 0° C.
Ether	6,000 atmospheres.
Benzene	11,000 ,,
Water	20,000

With such forces operating in the liquid state, and, to a lesser extent, in highly-compressed gases, it will be readily appreciated that the so-called correction term π, or $\frac{a}{V^2}$ in van der Waals' equation, is really far more important than the observed pressure P itself.

The Solid State of Matter.—The characteristic feature which distinguishes the solid state from the liquid and gaseous is the fact that a solid possesses a crystalline form. This means that the ultimate units or particles are not free to move about as they are in liquids, and still more in gases. They are, as it were, anchored to certain positions, and vibrate with respect to these positions. It is only thus that we can account for permanent crystalline form. For many years practically nothing was known about the solid state. Quite recently, however, very great progress has been made in the investigation of the structure of crystals, chiefly as a result of the work of W. L. and W. H. Bragg by means of X rays. It is impossible to enter upon this subject, but the striking conclusion which has been arrived at may at least be stated. It has been found that in crystals, such as crystals of sodium chloride, the ultimate individual or unit of matter composing such crystals is the atom and not the molecule. In fact, in crystals the concept of the molecule ceases to have meaning. Each sodium atom, for example, in sodium chloride, is equally attached to six chlorine atoms, and the same is true for each chlorine molecule. The structure is a lattice, or framework, with alternate atoms of sodium and chlorine at the corners. Naturally, not all crystals are equally simple; but even quite complicated crystals have been examined, and their structure elucidated, by the X-ray method. In view of this structure, which does not appear to have anything in common with liquids or gases, it is scarcely to be expected that anything like Andrews' idea of continuity of state can exist between the solid and the liquid state. That is, there is no such thing as an equation, analogous to van der Waals' equation, for example, which can be applied to, or, indeed, could have meaning for, the solid state. For further information regarding the structure of the solid state, the reader is referred to the book upon the subject written by the investigators named.

CHAPTER III.

MOLECULAR MAGNITUDES AND THE STRUCTURE OF THE ATOM.

IN the preceding chapter we have had occasion to deal with a certain number of molecular magnitudes, such as the velocity, momentum, and kinetic energy of molecules, in order to account for the laws of Boyle, Gay-Lussac and Charles, and the hypothesis of Avogadro. It is of interest, therefore, to obtain an idea of the numerical values of some molecular magnitudes which are characteristic of molecules of different gases.

The first point with which we deal is the number of actual molecules in one grammolecule of any gas, the so-called Avogadro Constant or Number, usually denoted by the symbol N. As already explained, the grammolecule or grammolecular weight of any gas denotes a mass which is a *multiple* of the true mass of the actual molecule.* The actual

---

* It may be pointed out that the grammolecular weight of any substance is a purely arithmetical quantity, not a physical or chemical one. It dates back to the time when no one had any idea of the true mass of a single actual molecule. It was

mass of any molecule is an exceedingly small quantity, a very small fraction of a gram. On the other hand, the grammolecular weight is usually a large quantity—*e.g.*, 2 for hydrogen, 28 for nitrogen, and 32 for oxygen. It follows, therefore, that the *multiple* or factor which connects the true mass of a molecule with the grammolecular weight is an exceedingly large number. This number is known as the Avogadro Constant or Number, N, and denotes the number of times the actual molecular weight or mass of a molecule can be divided into the grammolecular weight, or, what is the same thing, N denotes the number of actual molecules in one grammolecule. N has the same value for all gases.

The concept of a molecule as being an exceedingly minute discrete particle is, of course, quite old, and the so-called molecular theory of matter had been employed for a century as a working hypothesis before the real existence of molecules was demonstrated as an experimental fact. The demonstration consists in a determination of the Avogadro Number N. Within quite recent years this has been effected

arbitrarily assumed that the gramatomic weight, or more briefly the atomic weight of the hydrogen *atom*, was unity—that is, one gram. The molecule of hydrogen was known to contain two atoms. Hence the grammolecular weight of molecular hydrogen (ordinary hydrogen gas), is two grams, that of oxygen being 32, and so on. It was known, of course, that all these numbers were correct in a relative sense, and that a determination of N would render it possible to write down the actual masses of molecules of which we only knew previously the grammolecular weight. Thus the mass of the oxygen molecule is 16 times that of the hydrogen molecule, and so on.

by Perrin in France, and a little later by Millikan in America.

The phenomenon investigated by Perrin—a phenomenon which at first sight appears to have very little to do with the existence of molecules—was the well-known Brownian movement of small particles suspended in a liquid medium. If, for example, we examine by means of a high-power microscope a liquid in which very fine particles are suspended, such as gamboge in water, or the still finer particles of colloidal solutions, which can be rendered visible indirectly by means of the ultramicroscope, we find that these particles are in a state of rapid and irregular motion. Of course such particles, although very small, are considerably larger than single molecules, for molecules are too small to be detected directly even by the most efficient ultramicroscope. Investigations of Ramsay, and still more those of Gouy, served to demonstrate that this Brownian movement, exhibited by small particles, has its origin in the bombardment of the small particles by the invisible molecules of the medium in which the particles are suspended. Perrin succeeded in putting this conclusion to a quantitative test. He showed, for example, that the suspended particles distributed themselves in such a manner that their concentration—*i.e.*, their number per c.c.—near the bottom of a vertical cylinder, which contained the suspension, was considerably greater than their concentration at the top; and that, in fact, the concentration varied with the height according to a simple law—the same law which governs the density of the earth's atmosphere at different heights. Having obtained the necessary data in a fairly exact form, Perrin proceeded to apply certain theoretical considerations, into which we need not go, but about which there can be no doubt—considerations which necessarily follow from the assumption that the Brownian movement is really due to bombardment of the particles by the molecules of the medium. In this way Perrin succeeded in establishing a relation between the concentration gradient—i.e., the change in concentration of the particles with the height—of the gamboge particles and the quantity N. The value thus obtained for N agreed so well with certain approximate determinations which had been previously carried out by quite different methods that there could be no reasonable doubt but that Brownian movement was actually due to bombardment of the particles by the molecules of the medium. Hence, although we have not yet succeeded in seeing a molecule, we have actually before our eyes an irregular motion, analogous to molecular motion itself, and directly caused by the movements of the molecules which compose the liquid medium, water. Later on, Millikan improved the technique of such experiments by following the movement of a tiny droplet of oil suspended in air, and has in this way arrived at a very exact value for the Avogadro

Number. We may take this Number, N, to be 6.1×10^{23}.

If we pause to consider the enormous size of this number—a number which is known at the same time with a considerable degree of precision—we cannot but be struck by the great advance which such researches have effected. The above number means that there are 6.1×10^{23} molecules in one grammolecular weight of any gas—*e.g.*, in 2 grams of hydrogen, or 28 grams of nitrogen, or 32 grams of oxygen, and so on. Now we have pointed out in the preceding chapter that one grammolecule of any gas, at zero centigrade and at one atmosphere, occupies 22.3 litres. Hence the actual number of molecules in this volume is 6.1×10^{23}. Under the same conditions as regards temperature and pressure, the number of molecules in one c.c. of any gas is very nearly 2.8×10^{19}. Further, from the value of N and the value of the grammolecular weight, it is easy to calculate the true mass of a single molecule of any gas. Thus the grammolecular weight of hydrogen is 2 grams—*i.e.*, 2 grams of hydrogen contain 6.1×10^{23} molecules. Hence the mass of a single molecule of hydrogen is 3.2×10^{-24} grams. The true mass of a single oxygen molecule is 16 times the above quantity, and so on. These are extremely minute quantities.

The next molecular magnitude which we shall consider is the *diameter* of a molecule. Knowing the number of molecules in one c.c.

of gas, and likewise knowing the viscosity of the gas—a physical property sometimes called the internal friction—a simple relation enables us to calculate the size of any molecule—*i.e.*, the diameter. The following table contains a few values obtained in this way by Sutherland :—

Gas.	Diameter of Molecule.
Hydrogen	2.17×10^{-8} cm.
Oxygen	2.71×10^{-8} ,,
Nitrogen	2.97×10^{-8} ,,
Chlorine	3.74×10^{-8} ,,

The diameter of a molecule is therefore of the order of one hundred-millionth of a centimetre. If we imagine a number of hydrogen molecules placed in a row, and just touching one another, it would require fifty million of them to make a row one cm. in length. To give a more tangible, though less exact, idea of the minute size of a molecule, we may recall Lord Kelvin's well-known illustration—*viz.*, if we imagine a small drop of water magnified to the size of the earth, the molecules would then be visible, their size being somewhere between that of a cricket ball and that of small shot. The picture which we have in mind is that a molecule is a tiny sphere which can collide with and rebound from other molecular spheres.

The next quantity with which we have to deal is that known as the *mean free path* of a molecule in the gaseous state—*i.e.*, the mean or average distance which a molecule traverses between two consecutive collisions with other molecules. In the following table are given the values of the mean free paths of a few gases at zero centigrade and one atmosphere pressure, together with the values of the collision frequency—*i.e.*, the number of collisions which on the average any single molecule experiences in one second. This latter quantity is exceedingly large, one hydrogen molecule, for example, colliding with its neighbours some nine thousand million times per second.

Gas.	Mean Free Path in Cms.	Collision Frequency.
Hydrogen .	0.0000182	9.28×10^9
Oxygen .	0.0000094	4.82×10^9
Nitrogen .	0.0000100	4.28×10^9

It will be observed that at moderate pressure the mean free path of a molecule is very much greater than the actual diameter of a molecule —in fact, several hundred times as great. The mean free path is not to be confused with the average distance of molecules apart. Thus in the case of any gas at zero centigrade and at one-atmosphere pressure we now know that 6.1×10^{23} molecules occupy a space of 22.3 litres. That is, one molecule "occupies" a space of 4×10^{-20} cubic centimetres. If we look upon this little volume as a spherical space, its diameter is of the order 10^{-6} to 10^{-7} cm. The actual diameter of a molecule is of the order 10^{-8} cm., so that the actual diameter of a molecule is about one-tenth of the diameter of the space allotted to the molecule, or " occupied " by it under the given conditions of temperature and pressure. Expressing the same thing in a different way, we can say that the space allotted to a molecule under the above conditions, as regards temperature and pressure, is about one thousand times as great as the actual volume of the molecule. It is justifiable therefore, in any elementary treatment of the gaseous state, to neglect the actual volume of the molecules compared with the total volume of the gas. We have already made this assumption in the preceding chapter in deducing Boyle's law.

We now come to the question of the value of the average velocity of a molecule of a gas at a given temperature. It will be recalled that we have already deduced the following relation between the pressure of a gas and the density of the gas :—

$$P = \frac{\rho}{3} u^2$$

where u is taken to be the average velocity of a molecule. At a pressure of one atmosphere —i.e., $P = 1$, and at zero centigrade—the density ρ of hydrogen gas is shown by experiment to be 0.00009 grams per cubic cm. To obtain u in cms. per second, it is necessary to express the pressure also in the c-g-s. system of units. On this system it is known that one atmosphere is the same thing as 1.016×10^6 dynes per square cm. Hence, substituting these numerical values for P and ρ in the above equation, we find that

$$u = 1.8 \times 10^5 \text{ cms. per second.}$$

In words, this is a velocity of nearly twenty thousand cms. per second—a very considerable quantity. Other gases of greater density, such as oxygen, etc., have correspondingly smaller velocities. Knowing the mass of a molecule and its average velocity at a given temperature, we can easily calculate its average kinetic energy, for the kinetic energy is given by the expression $\frac{1}{2}mu^2$, where m is the mass of a molecule.

In the foregoing paragraphs we have been dealing with some of the more important characteristics of molecules *qua* molecules. The next point for consideration is the structure of the molecule.

In general, molecules contain two or more atoms. The exceptions to this statement are the rare gases of the atmosphere and the vapours of metals, which are monatomic. Chemical reactions in the majority of cases are essentially atomic reactions in the sense that they involve the transfer of an atom or group of atoms from one molecule to another, or the building up of quite new molecular structures by the assembling of atoms obtained from other molecules. This will be considered in its proper place. For the present we are concerned with the structure of the molecule in general. The question really resolves itself into a question of the structure of the atom, for the molecule is formed by the union of two or more atoms.

The Structure of the Atom. — All modern theories of the structure of the atom regard an atom as consisting of still smaller units of matter, each carrying an electrical charge. Before entering on the problem of atomic structure, it is necessary to recall one or two points about electricity itself which are of importance for our present purpose.

The first evidence that electricity could be transported through a solid was obtained when an electrical current was propagated through a wire by joining the two ends of the wire to the poles of a battery. The modern view of an electrical current is that it consists of a stream of very tiny particles called *electrons*, or corpuscles, each of which carries an electric charge, and each of which is very much smaller than an atom. The sign of the electric charge carried by an electron has been shown to be negative. These electrically-charged electrons existed in the wire all the time, their movement from one end to the other being brought about by joining the wire to the poles of the battery. Since the electrons were thus present in the wire, they must have formed part of the atoms of which the wire was composed. But the wire as a whole is electrically neutral. Hence there must be just as much positive electricity present in the wire—*i.e.*, in the matter composing the wire—as there is negative electricity. Positive electricity is now regarded as merely indicating absence of a certain amount of negative electricity. Such facts as these lead us to think of electricity as something which can be taken from or added to an atom or molecule—the latter being simply two or more atoms joined together. It has been shown, however, that we cannot add or take away any amount of electricity we please. There is a certain limiting quantity, a natural unit of electricity. This is the quantity which is carried by the electron. In the electron, therefore, matter and electricity have become one.

The discovery of electrons was first made by Sir J. J. Thomson. It was shown that electrons constitute the so-called cathode rays, a stream of very fine particles which is produced when an electric discharge is passed through a gas at low pressure. In addition to the cathode rays, we likewise meet with the production of electrons from the radioactive elements, from the alkali metals and alkaline earth metals, and in general from any material which is suitably excited—*e.g.*, by exposure of the material to ultraviolet light. It is natural therefore to regard the electron, with Sir J. J. Thomson, as " one of the bricks of which atoms are built up."

A very accurate measurement, carried out by Millikan, of the charge on an electron gives the value 4.774×10^{-10} electrostatic units. This value depends, as a matter of fact, on the determination of the Avogadro Number to which we have already referred. It is a striking example of the close interrelations which exist between fundamental units and quantities in molecular physics.

As already stated, the charge on the electron is the smallest quantity of electricity capable of independent existence. The primary quality of matter is mass, and an electron is the smallest mass of material conceivable. No smaller mass has ever been discovered, although it has been carefully looked for. It is known that the mass of an electron is *one-eighteen-hundredth part* of the mass of a single atom of hydrogen. That is, the mass of an electron is 5×10^{-28} gram. This quantity is so very small that obviously one electron could be detached from an atom without apparently altering the mass of the atom.

Atoms as usually met with are electrically neutral. They cannot consist therefore of electrons alone, for these would give rise to negatively charged systems; and, further, an assemblage of electrons, owing to their mutual repulsions, could not possibly produce a stable arrangement. An atom must contain at the same time some positive electricity, equal in amount to the total negative electricity which it contains. At present we know relatively little about the positive electricity in an atom. Simply for the purpose of giving stability to an atom, and making it electrically neutral, Sir J. J. Thomson assumed that an atom consisted of a sphere of positive electricity of uniform density, throughout which the electrons were distributed. With this assumption Thomson went on to consider the question of the distribution of the electrons in such a sphere which would correspond to a position of stability or permanence. The problem is to find how the electrons will distribute themselves in a sphere of positive electricity when successive additions of electrons are imagined to be made to the system. To simplify the treatment, Sir J. J. Thomson considered the distribution of electrons in one plane only—*viz.*, the plane through the centre of the atom. The results, though not complete, were very striking, and showed that the first important step in elucidating the structure of the atom had been made. We shall follow out a few of the simpler steps in this imaginary process of building up an atom model.

If one electron be added to an " empty " positive sphere, it will obviously go to the centre and remain there. If a second electron be added, the two will take up positions equidistant from the centre, their distance apart being identical with the radius of the sphere. Three electrons will distribute themselves at the apices of an equilateral triangle. Four electrons were shown to be incapable of stable equilibrium in one plane.   With five electrons the equilibrium state reached corresponds to a regular pentagon, one electron being at each corner.   Six electrons, however, did not form a stable hexagonal ring.   Instead, one electron goes to the centre, the remaining five forming a pentagon.   This is called a two-ring system.   Similarly, eleven electrons distribute themselves in such a way that two form the inner ring and nine the outer.   With successive additions, one finds that the two-ring system becomes unstable, and a three-ring system makes its appearance.

This occurs when seventeen electrons are present, and the three-ring system persists until we reach thirty-two electrons, when a four-ring system appears.   In this way we can build up more and more complex atomic models.   It will be evident at once that, as this process goes on, we have *periodic changes* in the type of distribution involved.   Thus, the atom containing one electron has this electron at the centre.   The atom containing six electrons has again one at the centre, the remainder forming an outer ring.   The atom with seventeen electrons is three-ring—one electron being in the centre, then a ring of five, and lastly a ring of eleven.   The atom containing thirty-two electrons is four-ring— one electron in the centre, then a ring of five, then a ring of eleven, and lastly a ring of fifteen.   This recurrence of the same type of structure—in the case considered, the recurrence is that of one central electron—at various intervals as the atomic weight increases, leads us to expect a recurrence of similar properties. This is the analogue of the Periodic Law. A discussion of the Periodic Law will be found in the book on *Inorganic Chemistry* in this series.

Although this demonstration of periodicity in properties was a very great step, the question of the distribution of the positive charge was still left quite unsettled. More recently, Rutherford has developed an atomic model, in which the positive charge resides on a *nucleus* at the centre of the atom, this nucleus being itself small compared with the dimensions of the atom as a whole. The Rutherford atom somewhat resembles a planet surrounded by satellites. We can speak of the nucleus, and the " atmosphere " of electrons rotating around it, as constituting the atom. In spite of its minute size, the nucleus is mainly responsible for the mass of the atom. It is not easy to enter into the reasoning which suggested this structure to Rutherford. Suffice it to say that the concept is based essentially upon the behaviour of an atom when an alpha particle from a radioactive substance collides with the atom. What is particularly important to point out, however, is that the Rutherford atom as it stands is unstable. An attempt to give it stability has been made by Bohr by introducing new assumptions which we shall not attempt to discuss here. There are one or two points still to be mentioned in connection with this atom-model. It must not be concluded that the central positive nucleus, although positively charged, is devoid of electrons. The nucleus does contain a number of electrons which are held extremely tenaciously—much more firmly, indeed, than the outer " atmosphere " of electrons. It is these outer satellite electrons which give rise to the chemical reactivity of the atom, its valency, etc.; and it is upon these outer electrons that ultraviolet light acts when an electron is detached from an atom by the influence of the light. As regards the number of these outer electrons, we are already in a position to say that the number is very nearly one-half of the ordinary atomic weight of the atom concerned. In the case of nitrogen, for example, the number of electrons is seven, in oxygen eight, in sodium eleven, and so on. In the case of hydrogen, the neutral hydrogen *atom* consists of a positive nucleus carrying unit positive charge round which a single electron rotates. A *molécule* of hydrogen contains two atoms. According to Bohr, the hydrogen molecule consists of two hydrogen nuclei, each carrying unit positive electric charge ; and between these nuclei the two electrons rotate in an orbit, the plane of which is at right angles to the imaginary line joining the two nuclei. This system can be shown on certain assumptions to be a stable one, provided the nuclei as they vibrate to and fro do not exceed certain limits of amplitude. There is a similar limitation to the diameter of the orbit of the electrons. A molecule of chlorine contains two atoms of chlorine, each of which consists of a nucleus and an atmosphere of seventeen electrons. Since the atom of chlorine is univalent, just as the hydrogen atom is, there must be one valency electron present in each atom of chlorine, whose special function it is to keep the two atoms joined together in the chlorine molecule. That is, we think of a chlorine molecule as consisting of two nuclei, each nucleus surrounded by rings of electrons, sixteen electrons per nucleus—these electrons being situated relatively close to the nucleus—and further out a ring of two electrons rotating in an orbit just as in the case of the hydrogen molecule. These two electrons are the valency electrons in the chlorine molecule, one of them belonging to each of the two atoms. If, now, a hydrogen molecule collides with a chlorine molecule, and if sufficient displacement of the parts be produced, it is conceivable that each molecule is broken in such a way that a chlorine nucleus, with its attendant sixteen electrons, unites with a hydrogen nucleus by means of the two rotating valency electrons, so as to give rise to an unsymmetrical molecule containing one chlorine nucleus, with its sixteen rotating electrons close to it, and one hydrogen nucleus, with a ring of two electrons rotating in a plane perpendicular to the line joining the nuclei. This new kind of molecule indicates that a chemical reaction has taken place between the hydrogen gas and the chlorine gas, with the production of a new substance, hydrogen chloride (HCl). This single example may serve to show that *chemical change involves a rearrangement of electrons and nuclei.* Owing to lack of knowledge we are unable to pursue this mode of treatment of chemical change much further. Other methods of attack are, however, available, and these will be considered in the succeeding chapters.

CHAPTER IV.

CHEMICAL REACTIONS IN GASES.

DEALING with the problem of chemical reactions between different kinds of matter, the question was early raised as to whether there was any connection between the *extent* of a reaction and the quantities of the substances participating. There is a further question as to whether there is any connection between the *speed* or *rate* or *velocity* of a reaction and the quantity of material employed.

In attempting to answer these two questions we shall take the simplest possible physical conditions—*viz.*, a reaction in the gaseous state. Let us think, first of all, of a mixture of oxygen and hydrogen gases. Under certain conditions these two substances react together to form water vapour, the reaction being represented by the chemical equation—

$$2H_2 + O_2 = 2H_2O,$$

where $2H_2$ stands for two molecules of hydrogen, O_2 for one molecule of oxygen, and $2H_2O$ for two molecules of water vapour or steam. As a rule, this reaction proceeds exceedingly rapidly, in fact explosively. Other reactions are known, however, in which the rate of combination is sufficiently slow to be measured, say, by analysing the system at various intervals of time. Still, other reactions are so very slow as scarcely to be measurable at all.

In spite of the fact that hydrogen and oxygen usually react together immeasurably rapidly, it has nevertheless been shown that the union is never really complete. Very careful experiments—as a matter of fact, at high temperatures—have shown that there is always a minute amount of hydrogen and oxygen left uncombined; and this absence of completeness is characteristic of all reactions in a gaseous, or in general in a homogeneous, state. At room temperature, indeed, in the case of the reaction between hydrogen and oxygen, the amount of uncombined gases is far too small to be detected; but it can be detected at higher temperatures, because there is a greater amount left uncombined at higher temperatures. The explanation of this lack of completeness is that the molecules of water vapour themselves are capable of reacting with one another to produce hydrogen and oxygen molecules. Such reactions which do not complete themselves are called *reversible reactions*. The position at which the reaction apparently ceases is called the *equilibrium point*. In some cases the equilibrium lies far over to one side; so far over, in fact, that the reaction appears to be complete. This is the case with the union of hydrogen and oxygen at ordinary temperatures. In other cases the equilibrium point is quite easily observable. Thus the decomposition of nitrogen tetroxide into nitrogen dioxide, which is represented by the chemical equation thus—

$$N_2O_4 = 2NO_2,$$

is a reaction which comes to a stop, even at ordinary temperatures, at a point at which quite sensible amounts of both N_2O_4 and NO_2 are present together. The position of the equilibrium point measures the *extent* of a reaction.

In all these cases the equilibrium point, although it appears to correspond to a stoppage of the reaction, does not in reality mean such a cessation. Even *at* the equilibrium point the N_2O_4 molecules are continuing to decompose into NO_2 molecules; but at the same time just as many NO_2 molecules are combining to form N_2O_4 molecules. We have, therefore, a balance of two reaction velocities or rates; and since they compensate one another, the system *as a whole* no longer changes with time. This kind of equilibrium, which is quite characteristic of chemical changes, is consequently a dynamic one, not a static one.

It is evident that the position of the equilibrium depends upon the relative velocities of the two opposing processes. To indicate that an equilibrium is set up, and that the equilibrium point depends upon the velocity of the opposing reactions, it is usual to write the chemical equation in the following way—

$$2H_2 + O_2 \rightleftarrows 2H_2O,$$
$$N_2O_4 \rightleftarrows 2NO_2,$$

the arrow denoting the direction of change. If now the change from left to right goes with a speed or rate or velocity which is much greater than that of the reverse process, it is obvious that the position of the equilibrium will lie far over to the right. This is the case with the union of hydrogen and oxygen, this reaction finally attaining an equilibrium which corresponds to a very large amount of water vapour compared with the quantities of hydrogen and oxygen.

Now there is a general principle which, though exceedingly simple, has been found to govern not only the rate of a chemical reaction, but likewise the position at which equilibrium is finally attained. This principle is known as the *Principle of Mass Action or the Principle of Active Mass*. As this lies at the basis of chemical processes in general, we must consider it in some detail.

First of all, let us take up the now familiar idea that a chemical reaction proceeds at a certain rate. By the rate in such a case we mean the number of grammolecules of a substance which decompose or react with some other substance per second. It differs, therefore, from the simple mechanical idea of rate ; but at the same time the term is not ill-chosen. When we examine a reaction—a reaction between gases, say—and measure the rate of the reaction, we encounter a striking fact, which is common to all chemical processes—*viz.,* the fact that the rate diminishes as the reaction proceeds. In fact, the rate starts off at a high value, and gradually decreases until the end point is attained. We can estimate the rate at different stages or time intervals by analysing the mixture in some way (either chemical or physical), and find how much of the original substance remains in existence. Naturally there will be less and less of this substance remaining as the time goes on. We can express this fact in the form of a graph (Fig. 4), in which the ordinate denotes the

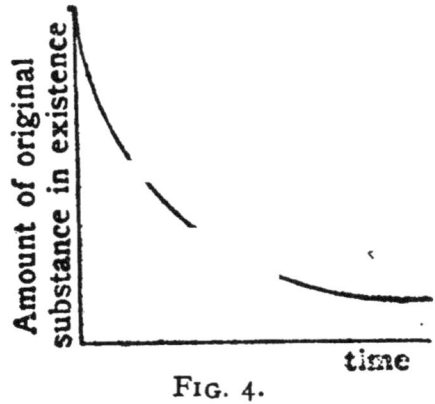

FIG. 4.

amount of the original substance left, and the abscissæ represent the time from the commencement of the reaction. The slope of this curve measures the rate of the reaction at any moment of time. It will be seen that the curve starts off steeply—*i.e.*, the initial rate is very high. As the reaction proceeds the rate falls off, the curve becoming less and less steep, so that at a later moment of time the rate is now small. The rate of a reaction is, therefore, not a fixed or constant quantity for the reaction, even if—as is assumed—the temperature of the system is kept constant throughout the experiment. We naturally ask why the rate should fall off in this manner, and the answer is given by the principle of mass action, which may be stated as follows: *The reactivity of a substance—i.e., the rate at which it reacts or decomposes—is directly proportional to the concentration of the substance at the moment considered.*

It is clear that this principle expresses the experimental fact to which we have just alluded. At the commencement of the reaction the concentration of the gas—*i.e.*, the number of grammolecules of the substance present in one litre—is large, and consequently the rate of decomposition is large. Later on in the process the concentration of the original substance gas is less; and hence, on the basis of the principle, the rate should be correspondingly less. As long as any of the original gas is present, it must still be decomposing in proportion to its concentration. That is, at the equilibrium point the reaction must still be going on, although it is there just compensated—as far as net observable change is concerned—by the reverse process. Naturally the reverse process also occurs in accordance with the principle of mass action. During the reaction the products formed as a result of the reaction are collecting—*i.e.*, their concentration is increasing, and hence the rate at

which they unite to re-form the original substance is increasing with the time, until this
rate just balances the rate of the original reaction.   When this balance is attained the equilibrium point is reached.

For the sake of simplicity let us think of
what is called a *unimolecular* change—*i.e.*,
the spontaneous decomposition of molecules
one at a time.   Such changes are represented
by the act of decomposition of N_2O_4 into NO_2,
and by the dissociation of a gas such as chlorine
into its atoms, thus : $Cl_2 \rightarrow 2Cl$.   (It happens
that in neither of these cases is it possible to
measure the rate of dissociation, for it is too
great, and the equilibrium point is attained
in a fraction of a second.   This, however, does
not affect the point under discussion ; both
reactions are typical unimolecular ones.   For
our present purpose we are thinking of such
a reaction as occurring at a measurable rate.)

Now it would obviously be a very useful
thing to have some numerical quantity which
would serve to characterise any given reaction in order to compare different reactions
with each other.   We have seen that the rate
of a reaction cannot be employed for this purpose, because of the fact that the rate or velocity varies with the time.   By introducing
the principle of mass action, however, as we
have already done, a means is given us where-
by we may characterise a reaction by making
use of what is known as the *velocity constant* of
a reaction.   This may be readily explained.

According to the principle of mass action as applied to a simple reaction, such as the dissociation of a gas, we have seen that the rate is proportional to the concentration of the undecomposed gas at the moment considered. Instead of using the term " is proportional to," it is always permissible to introduce the sign of equality, and insert at the same time a numerical factor or constant of proportionality, which we shall denote by the symbol k_1. Hence the principle of mass action may be stated thus :

Rate of the reaction = $k_1 \times$ concentration of decomposing substance,

Or, number of grammolecules decomposed per sec. = $k_1 \times$ concentration of decomposing substance.

Let us denote concentration in general by C. This does not represent a fixed quantity ; it denotes the value of the concentration, whatever that may happen to be, at a given moment of time. We can thus write the principle of mass action in the very brief form—

Rate = $k_1 C$.

This proportionality factor k_1 is called the velocity constant of the reaction ; k_1 is characterised by the fact that it is a numerical *constant* at constant temperature, independent of the time. Thus, although the rate falls off as C diminishes, k_1 nevertheless maintains a fixed value. The reaction may therefore be characterised by the value of k_1, although it cannot be characterised by the rate itself. Naturally the value of k_1 can be determined from a series of measurements of the rate at known time intervals.

The simplest kind of reversible reaction would be represented by two opposed unimolecular processes, thus—

$$A \rightleftarrows B,$$

where A and B are, in effect, isomeric substances. It is rather a hypothetical case as far as gases are concerned, but it will serve as a guide for similar reactions which are known to occur in solutions. Let us suppose that the velocity constant of change of A—*i.e.*, the process from left to right—is k_1, and that k_2 is the corresponding velocity constant of the process from right to left—*viz.*, the change of B into A. Further, let us suppose that the equilibrium point has been reached —*i.e.*, the stage at which there is no longer any *net* change in the system. We have seen that this realised when the rate of the reaction from left to right balances the rate of reaction from right to left. Suppose that the system in this state is analysed, the equilibrium concentration of A—*i.e.*, the concentration of A at the equilibrium point—being denoted by C_A grammolecules per litre, the corresponding quantity for the substance B being denoted by C_B. It follows that the rate of disappearance of $A = k_1 C_A$, whilst at the same moment the rate of disappearance of $B = k_2 C_B$.

Now when B disappears, or changes, it obviously gives rise to A. Further, when the equilibrium stage has been reached, it follows that the rate of disappearance of A is just equal to its rate of appearance or formation from B. That is, the two rates mentioned above are equal at the equilibrium point. Hence as a criterion of the equilibrium point we have—

$$k_1 C_A = k_2 C_B.$$

Note that we do not say that the velocity *constants*, k_1 and k_2, are equal. This in general is never the case. What we do say is that the rate itself—which is the product of the velocity constant by the concentration of the substance—of the reaction in one direction is equal to the opposing rate.

The above relation may be written in the slightly different form—

$$\frac{C_B}{C_A} = \frac{k_1}{k_2}.$$

Now $\dfrac{k_1}{k_2}$ is simply the ratio of two constants.

Such a ratio must itself represent a constant quantity. It is usual to denote this ratio by K, where K defines the equilibrium point, and is known as the *Equilibrium Constant*. The equilibrium constant of a reaction is, therefore, simply the ratio of the two opposing velocity constants. We can, therefore, rewrite the expression which characterises the point of equilibrium in the following way—

$$\frac{C_B}{C_A} = K,$$

where C_A and C_B denote the equilibrium concentrations.

This expression means that, whether we start with a system consisting entirely of molecules of the substance A, or entirely of B, a chemical change will occur in such a sense as finally to give both A and B in such proportions that the ratio of their concentrations has the value K.   This is the sense in which K is a constant at all.   The concept of the equilibrium constant is fully as important as, if not more important than, the concept of velocity constant.   That there should be such a quantity as K to characterise equilibrium depends directly upon the application of the principle of mass action.   The numerical value of K will naturally vary with the chemical nature of the system or substances reacting.   Its value also varies with the temperature.

Numerous reactions in gases and in solutions have been examined, and it has been found that the equilibrium finally attained is, indeed, governed by an equilibrium constant, which in turn is the ratio of the two opposing velocity constants.   That is, the mass action concept of equilibrium and reaction velocity is fully borne out by experiment.   The velocity or rate constant serves to characterise the speed or rate of a reaction ;   the equilibrium constant serves to characterise the extent to which the reaction will proceed.

Hitherto we have been restricting ourselves, for the sake of simplicity, to unimolecular reactions alone. Other kinds of reactions— multimolecular reactions as they are called— are likewise known. Of these the most important are the so-called *bimolecular reactions* —*i.e.*, reactions which result from the interaction of two molecules which may be chemically the same or different. A typical bimolecular reaction is represented by—

$$A+B=D+E,$$

where A, B, D, and E are four different gases. One molecule of the substance A must collide with one molecule of the substance B before D and E can be formed. If the reaction is reversible, then one molecule of D can react with one molecule of E to re-form A and B. The question is: How are we to express the rate of either of these processes on the basis of the principle of mass action? We are considering, in the first place, the reaction from left to right—*i.e.*, the rate of union of A and B. Further, let us think of some moment of time *before* the equilibrium has been reached. On applying the principle of mass action we would say that the rate at which A reacts is directly proportional to its concentration, and the same is true of B. This is expressed by writing—

Rate of combination of A and $B=k_1 C_A \times C_B$. Note that the velocity or rate of union is proportional to the *product* of the concentrations of A and B at the moment considered. k_1 stands as before for the velocity constant of the bimolecular process. To see why the pro-

duct of the concentration of each of the reacting species comes in (and not the sum of
the concentrations, for example), let us take
an actual case—*viz.*, the union of hydrogen
(H_2) with chlorine (Cl_2), both in the gaseous
state to form gaseous hydrogen chloride (HCl).

Suppose that we have in a given space equal
numbers of hydrogen molecules and chlorine
molecules. A certain number of collisions in
which a hydrogen molecule strikes a chlorine
molecule will occur in one second, and in some
of these collisions combination will take place.
Now suppose the concentration of the hydrogen
to be doubled: then *twice* as many hydrogen
molecules will collide with a chlorine molecule
per second, and *twice* as much hydrogen
chloride will be formed per second. If the
concentration of chlorine alone be doubled,
the same effect will be produced, so that, if
both concentrations be *doubled*—say by com-
pressing the gaseous mixture to one-half of its
original volume—then *four* times as many collisions will occur in unit volume in unit time,
and therefore the velocity of the reaction—
i.e., the amount of combination per second—
will be *quadrupled*. From this it will be seen
that the rate varies as the product of the concentration terms; for if we alter C_A to $2C_A$,
and C_B to $2C_B$, as in the above case, then
$k_1 \times$ product of the concentration terms is
$k_1 \times 2C_A \times 2C_B$, which is the same as $4k_1C_AC_B$,
which is four times the original rate, $k_1C_AC_B$.
The use of the product of the concentration

terms in all such cases has been fully borne out by experiment. Thus the rate at which D and E unite is given by $k_2 C_D C_E$, where k_2 is the velocity constant of the reaction from right to left.

Now let us consider the equilibrium point itself. Suppose that when this stage is reached the concentration of each of the four gases is represented by the symbol—

$$C_A, \ C_B, \ C_D, \ \text{and} \ C_E,$$

where the equilibrium concentration is written in heavy type. Then at the equilibrium point we have—

Rate of disappearance of A and B $= k_1 C_A C_B$.
Similarly, the

Rate of disappearance of D and E $= k_2 C_D C_E$.
But at the equilibrium point these two rates must be the same. Hence the equilibrium point is characterised by the relation—

$$k_1 C_A C_B = k_2 C_D C_E \ ;$$
$$\text{or,} \qquad \frac{C_D \times C_E}{C_A \times C_B} = \frac{k_1}{k_2} = K,$$

where K is the equilibrium constant.

The meaning of the above expression is, that if we start with any amounts of all four substances that we please (*e.g.*, we can start with nothing but A and B, or nothing but D and E, or we can start with some of all four) the system will always spontaneously adjust itself. That is, it will adjust its composition by means of a chemical change, so as to give such values to the concentrations of the gases that the left-hand side of the expression will attain a value which is independent of the initial conditions—*i.e.*, independent of the initial composition of the mixture. This has been fully borne out by experiment.

In the above case we have been considering the reaction between two different kinds of molecules, A and B, or D and E. Exactly the same reasoning applies even when the two reacting molecules are chemically the same. This is met with, for example, in the decomposition of hydrogen iodide gas (HI), which decomposes to give rise to hydrogen gas (H_2) and iodine vapour (I_2), according to the chemical equation—

$$2HI = H_2 + I_2.$$

A single molecule of HI cannot give rise to a molecule of H_2 or of I_2. It has been found by experiment that two molecules of HI combine together to give hydrogen and iodine. The reaction may be written in another way, which is equivalent, of course, to that already given, *viz.*,—

$$HI + HI = H_2 + I_2,$$

which is formally analogous to—

$$A + B = D + E.$$

At the equilibrium position we have, therefore—

Rate of disappearance of hydrogen iodide
$$= k_1 c_{HI} \times c_{HI} = k_1 c^2_{HI}.$$

Similarly, the
Rate of disappearance of H_2 and $I_2 = k_2 c_{H_2} c_{I_2}.$

Whence the equilibrium is defined by the relation—

$$\frac{c_{H_2} c_{I_2}}{c^2_{HI}} = \frac{k_1}{k_2} = K,$$

where K is the equilibrium constant. This particular reaction has been very carefully investigated, and it is proposed to give some numerical values in connection with it.

In the first place, in calculating K, it is, of course, essential to be certain that the equilibrium point has really been reached. To make sure of this, it is necessary to approach the equilibrium from both sides. In the present case we must begin with pure HI in the reaction vessel, adjust the temperature to a suitable value in order that the reaction may proceed with sufficient rapidity, and finally analyse the mixture so as to obtain values for the equilibrium concentrations of HI, H_2, and I_2. The same experiment is repeated, giving it a longer time, and again analysed. If the same analytical figures are obtained, we conclude that we have allowed sufficient time, even in the first run, for the reaction to come to an end. This is certainly very strong evidence that the equilibrium point has been attained, but it can be improved upon. We start with hydrogen and iodine together, and raise the temperature to the same value as before. Formation of HI now occurs. Again allow sufficient time to elapse, and finally analyse the mixture. If the equilibrium point has been really attained in both cases, the same value for K will be given by the analytical results.

We can represent the above procedure graphically, as in Fig. 5, in which the abscissæ denote time, and the ordinate denotes concentration of hydrogen iodide gas. On starting with pure (100 °/₀) HI, the system follows the upper line, which eventually becomes

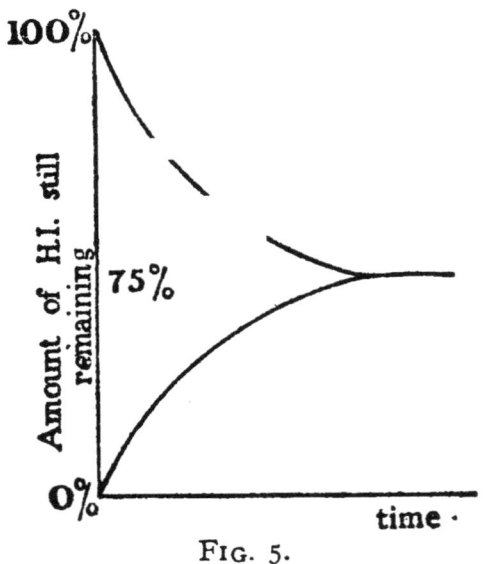

FIG. 5.

horizontal at the value 75 °/₀ HI at a particular temperature. On starting with a mixture of H_2 and I_2 (i.e., zero HI) the system changes, following the lower line, HI being formed until the same horizontal part is reached and the two lines have coalesced. Such behaviour is clear evidence that equilibrium has really been attained, the composition of the system at the equilibrium point being given by the position of the horizontal line. An actual operation may be briefly described.

A number of glass bulbs, each of about 15 c.c. capacity, were filled with HI gas at various pressures ($\frac{1}{2}$, 1, 1$\frac{1}{2}$, 2 atmospheres) at room temperature, and sealed off. They were then heated in baths at known temperatures between 100° C. and 518° C. for given intervals of time. The bulbs were then removed, cooled * as quickly as possible, and opened under caustic potash, which dissolved the hydrogen iodide still remaining, as well as the iodine formed in the process, and the hydrogen remained undissolved. From a knowledge of the volume of the bulbs, the mass of HI originally present was determined. By measuring the volume of hydrogen formed, we can calculate the quantity of iodine formed, for these two quantities must be chemically equivalent. The amount of HI undecomposed can be obtained by titration of the caustic. From these data we have all that we require to work out the equilibrium expression—

$$\frac{c_{H_2} c_{I_2}}{c_{HI}^2}.$$

In another series of experiments the bulbs

* This is known as chilling, or freezing, the equilibrium. The lowering in temperature is effected so quickly that the mixture really corresponds to the higher temperature in the experiment itself. The efficiency of the procedure depends upon the fact that on cooling the system the reaction velocity falls off very greatly, and there is practically no tendency to change at low temperatures. Hence analysis can be carried out conveniently.

were filled with known amounts of H_2 and I_2, the reaction allowed to take place at the same series of temperatures as in the first set of experiments, and analyses made of the systems after a certain time had elapsed. The curve of Fig. 5, already given, indicates the nature of the results obtained. The sloping parts of the curves indicate the state of affairs *before* equilibrium had been attained. The horizontal portion indicates that no further change is occurring with the time, the same amount of HI being found in both series of experiments. The value of

$$\frac{c_{H_2} c_{I_2}}{c^2_{HI}}$$

taken from the horizontal part of the curves gives the value of the equilibrium constant K.

The existence of an equilibrium point or position, and the value of the equilibrium constant, is an important physico-chemical characteristic of any reaction. An obvious technical application is this: no yield or production of a substance can exceed the equilibrium yield. Hence, if the equilibrium constant has been determined, it is possible to state the maximum amount of new material which can be produced under a given set of conditions in a technical process. Naturally the object of all such processes is to attain as great a yield as possible of some substance. By dealing with the problem in this physico-chemical way we are no longer following mere rule-of-thumb methods; for on obtaining a given yield in a technical process we can say whether this is the maximum attainable, or how far short of the maximum the process has fallen. In this way much time may be saved ultimately by thoroughly working out the velocity constants, if possible, and the equilibrium constants at various temperatures on a small scale before attempting to set up or modify a large-scale process.

To return to the hydrogen iodide decomposition, it may be of interest to give in the form of a table the observed amounts of HI which were actually formed when the reaction had ended in a set of experiments carried out at 357° C., in which the starting materials were hydrogen and iodine, and to compare these observed values with those calculated by means of the equilibrium mass action expression.

Reaction : $H_2+I_2=2HI$. Equilibrium conditions at 357° C.

Initial Quantity of Hydrogen.	Initial Quantity of Iodine.	Quantity of HI formed (observed).	Quantity of HI (calculated).
2.59	6.63	4.98	5.02
10.40	6.41	11.88	11.68
22.29	6.51	12.71	12.68
23.81	6.21	12.17	11.98

On comparing the values given in the last

two columns, it will be seen that they agree very satisfactorily. Such divergences as appear are due to unavoidable experimental error. The principle of mass action is conclusively demonstrated by this and numerous other reactions.

Only one point further may be mentioned. Temperature has an exceedingly marked effect upon the velocity constant of a reaction—*i.e.*, upon the quantities k_1 and k_2. In fact, if we raise the temperature from, say, 25° C. to 35° C.—*i.e.*, a rise of 10 degrees—the velocity constant is found to be about trebled, the actual multiples, which vary from reaction to reaction, lying between 2 and 4. Since the equilibrium constant is the ratio of two such quantities, it is obvious that K will also vary with the temperature, though to a much less extent. It is always necessary, therefore, to maintain the temperature constant during the determination of velocity constants and equilibrium constants. This variation of velocity and equilibrium constants with temperature is a fundamental factor in the control of large-scale technical processes.

CHAPTER V.

CHEMICAL REACTIONS IN SOLUTIONS.

WE have seen, in the case of chemical reactions in gases, that the principle of mass action is a very useful guide in determining to what extent a reaction will proceed, and the rate at which it occurs. We have now to see how far the same principle will assist us in dealing with chemical reactions between dissolved substances—that is, substances, either liquid or solid, which have been dissolved in a liquid medium. The most usual liquid medium is, of course, water.

Before dealing with the question of the applicability of the law of mass action to reactions under these conditions, it is necessary to consider briefly some of the more outstanding characteristics of solutions, quite apart from any chemical change which may occur in them.

Undoubtedly the most striking phenomenon met with in connection with solutions is that known as *osmotic pressure*. It appears as though the dissolved substance, or *solute*, as it is called, which is distributed throughout the medium, giving rise to a more or less dilute solution, acts like a gas in respect of the phenomenon of osmotic pressure. The phenomenon will be best understood by describing the arrangement for observing and measuring it.

Let us suppose that we have a solution (Fig. 6), say, of cane sugar in water, placed

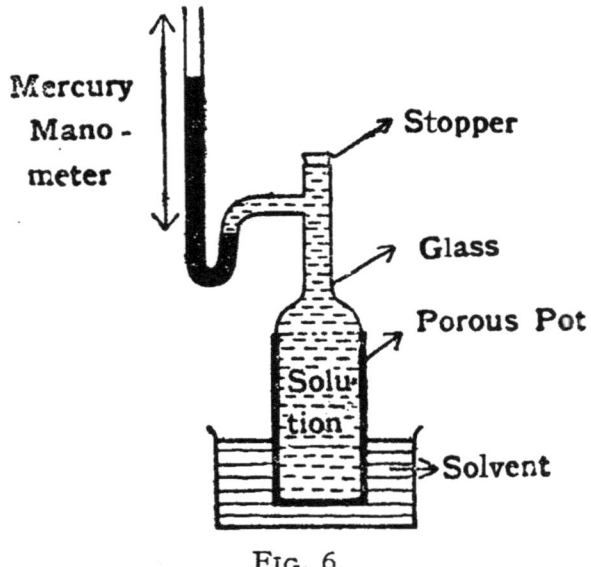

Fig. 6.

inside a porous earthenware pot, in the walls of which some copper ferrocyanide has been previously precipitated, so as to clog the pores sufficiently. The precipitate is supported in the walls to form a continuous membrane. This membrane has the property of allowing the solvent, water, to pass freely through it, but it completely prevents any passage of the dissolved sugar molecules. For this reason the membrane is said to be *semi-permeable*. Let us suppose that the porous pot, having the membrane in its walls and containing the sugar solution, is partly immersed in a wider vessel containing the pure solvent, water. If now the upper level of the solution be observed, it will be found to rise—*i.e.*, the solution expands in volume. If the solution fills the whole of the pot, and likewise the bent glass tube joined to it at the top, the solution being in contact with the mercury in the manometer tube shown in the figure, then the expansion of the solution will be observed by the rise of the column of mercury in the long limb. What has happened is, that water has passed from the outer vessel (which contains pure water only) *into* the solution. In order to oppose and prevent this entry of water, with the consequent expansion of volume of the solution, it is necessary to increase the head of mercury by adding mercury to the open limb—*i.e.*, to put on an opposing pressure. The pressure which has to be put on in this way in order to keep the solution at constant volume is a direct measure of what is called the osmotic pressure of the solution. The solution only exhibits this property of tending to expand when it is put into contact with pure solvent in the manner described. The solute seems to act like a gas in this respect—*i.e.*, in respect of expansion of the solution. One way of looking at the phenomenon is to ascribe it to bombardment of the inside of the walls of the porous pot by the molecules of the dissolved substance, just as a gas would do. Another and better way is to look upon the water in the outer vessel as being sucked into the solution owing to the presence of the solute in the inner vessel, the pot. The tendency to expand by drawing in water from the outside vessel exists in virtue of the property called osmotic pressure, or, rather, osmotic pressure is the result of this suction effect.

Now the remarkable thing about the osmotic pressure of a solution—which can be measured by means of addition of mercury, as already described—the remarkable thing is that the value of the osmotic pressure in the case of a dilute solution is identical with the pressure which would be exerted by a gas at the same concentration (grammolecules per litre) and temperature as that of the dissolved substance. Thus if we have one grammolecule of a gas at 0° centigrade enclosed in a volume of one litre, the gas exerts a pressure of 22.3 atmospheres. Similarly, if we make up a solution of cane sugar in water, so that there is one grammolecule of cane sugar in one litre of solution, and if we place this solution in the apparatus described, pure solvent will tend to be sucked in; and, in fact, it is necessary to put on an extra pressure on the solution by means of the mercury column, which is found to be very nearly 22.3 atmospheres, in order to keep the solution at its initial volume. That is, the

analogy between a gas and a dissolved substance is not only qualitative but quantitative.
The analogy is further borne out when we consider the influence of temperature. We have
already seen that the volume of a gas increases
in proportion to the absolute temperature of
the gas if the pressure be kept constant; or,
if the volume of the gas be kept constant, its
pressure rises in direct proportion to the absolute temperature. Similarly, if a solution be
kept at a fixed volume by applying pressure
by means of the mercury column—naturally
we do not apply any pressure to the surface of
the pure solvent—its osmotic pressure increases in direct proportion to the absolute
temperature. In fact, a dilute solution, or,
more properly speaking, the solute in a dilute
solution obeys the gas law—

$$PV = RT,$$

where P is the osmotic pressure of the solution,
V is the volume containing one grammolecule
of the solute, T is the absolute temperature, and
R is a constant which is identical in value with
the gas constant. The fact that R has the
same numerical value, reckoned per grammolecule of solute, as it possesses when reckoned
for one grammolecule of a gas, shows that
Avogadro's hypothesis also applies to the
solute in dilute solutions. That is, equimolecular solutions of different solutes contain the
same number of dissolved molecules.

When we think of osmotic pressure we always

think of it as measured in the manner described. Naturally the solution by itself, in a beaker, for example, does not obey the gas law, for, being a liquid, it is very incompressible. It is only when a solution is put into contact with the pure solvent, *via* a semipermeable membrane, that the osmotic pressure manifests itself. If the membrane happens to be faulty, and permits the dissolved substance to stream out into the pure solvent, we simply dilute the solution, and no osmotic pressure manifests itself.

We can regard the molecules of a dissolved substance as moving rapidly about, colliding very frequently with the molecules of the solvent and less frequently with one another, because there are few solute molecules compared with solvent molecules in any given space. To this extent a dissolved substance resembles a gas, but it would be going too far to say that the solute is identical with a gas. The discovery of the fact that a dissolved substance closely resembles a gas in giving rise to an osmotic pressure, which is quantitatively expressed by the gas law, was first made by van 't Hoff. Van 't Hoff found that not only cane sugar, but many other substances, when dissolved, exerted this pressure, and its magnitude could be calculated by the expression $P = \dfrac{RT}{V}$. Such substances were sugar, urea, acetamide, etc. The phenomenon is, therefore, a general property of solutions. It should

be observed that the substances mentioned,
and others which exert osmotic pressure in
agreement with the gas law, do not conduct
the electric current. They are therefore said
to be *non-electrolytes*.

Whilst dissolved non-electrolytes are known
to obey the gas law, van 't Hoff found that
there was likewise a large group of substances,
characterised by the fact that their solutions
in water do conduct electricity, which do not
obey the gas law in respect of their osmotic
pressure. The osmotic pressure of the substances of this group was always greater than
that calculated by the gas law. In other
words, they exert an osmotic pressure which
is greater than that exerted by a non-electrolyte at the same molecular concentration.
Since these substances, which exert an abnormally great osmotic pressure, are capable
of conducting electricity, they are called *electrolytes*. To this group belong ordinary acids,
such as hydrochloric, nitric, and sulphuric
acids; bases such as caustic soda, caustic
potash, and baryta; salts, such as potassium
chloride, sodium chloride, copper sulphate,
lead nitrate, and many others. Even salts of
organic acids, such as sodium acetate, conduct
electricity, and also exert an abnormally great
osmotic pressure.

In solutions of electrolytes it is now known
that the carriers of electricity are particles
identical in mass with the atoms of the substance. Some of these particles are charged

positively, some negatively. The solution as a whole is electrically neutral — *i.e.*, there must be just as many positively charged atoms or positive ions, as they are called, as there are negative ions. These ions are present all the time in the solution ; they are formed, in fact, in the act of dissolution of the solid salt in the water. The ions are free to move about throughout the solution. Their presence is shown when we cause them to move in a definite direction, by putting on an electric pressure or potential difference across the solution by means of electrodes inserted in the solution, the electrodes being connected to an external battery. It is then found that the positive ions move in one direction, the negative ions in the opposite, their joint effect constituting the passage of electricity through the solution. Each ion is attracted towards the electrode, which is electrically charged in an opposite sense to that of the ion itself. That is, positive ions move towards the negative electrode, or *cathode*, the negative ions moving towards the positive electrode or *anode*. The ions which move to the cathode are called *cations*, those which move towards the anode, *anions*. That is, cations are positively charged, anions negatively charged. Note that the anode is the electrode where the current generated from the battery enters the solution, and the cathode is where the current leaves the solution. The direction of current is in fact the same as that of the movement of the positive ions. When an ion reaches an electrode it gives up its charge of electricity to the electrode, the ion becoming discharged thereby. It is now simply a neutral atom. Several things may happen to this atom—*e.g.*, it may be deposited upon the electrode, as the discharged silver ions in electroplating, or it may go off in the form of gas, as happens when we electrolyse a solution of common salt (NaCl), obtaining gaseous chlorine at one of the electrodes—the anode, as a matter of fact—each molecule of gaseous chlorine consisting of two atoms of chlorine which have combined together after being discharged at the electrode. Or, finally, the discharged atom may react with the solvent, water, as happens in the case of sodium, giving rise to new chemical reactions, such as the production of hydrogen gas and the formation of caustic soda round the electrode where discharge of the ion first took place (the cathode, in the case of sodium ion). We may say in general that the ions of metals are positively charged—*i.e.*, they are cations; whilst ions from non-metallic elements, such as chlorine ion, bromine ion, iodine ion, sulphate ion, nitrate ion, are negatively charged —*i.e.*, they are anions.

It was Arrhenius who some thirty years ago first pointed out that the process which occurs when an electrolyte—*i.e.*, a salt, base, or acid—is dissolved in water, is a process of *electrolytic dissociation*, or, as it is frequently called, *ionisation*. In the case of potassium chloride dissolved in water the process can be represented thus—

$$KCl \rightarrow K^+ + Cl^-,$$

where KCl stands for a molecule of the undissociated potassium chloride—which is itself a non-conductor of electricity—K^+ denotes the positively charged potassium ion, and Cl^- denotes a negatively charged chlorine ion. Arrhenius was the first to insist upon the real and free existence of such ions, formed immediately the salt is dissolved. Further, Arrhenius showed that an equilibrium is set up between the ions and the undissociated molecules, the undissociated molecules continually dissociating into ions, whilst the ions at the same time unite to form undissociated molecules. This dynamic equilibrium is represented as before by means of the expression—

$$KCl \rightleftharpoons K^+ + Cl^-.$$

The corresponding reaction in the case of HCl is—

$$HCl \rightleftharpoons H^+ + Cl^-,$$

and, in the case of caustic soda—

$$NaOH \rightleftharpoons Na^+ + OH^-.$$

The most important contribution made by Arrhenius was, however, the demonstration that *the degree or extent of the electrolytic dissociation varies with the dilution of the electrolyte, increasing as the dilution increases up to a limit which corresponds practically (but not absolutely*) *to complete electrolytic dissociation.* By the term *dilution* is meant the volume of solution in which one grammolecule of the original substance is dissolved. Thus a decinormal solution of potassium chloride is one in which one grammolecule of KCl has been weighed out and dissolved in ten litres of water. Experiment has shown that in many cases—the so-called strong electrolytes—the dissociation is very nearly complete (about 99 per cent.) when the dilution is of the order of 1,000—*i.e.*, when one grammolecule of salt, acid, or base has been dissolved in 1,000 litres of water.

It will be observed that when *one* molecule suffers electrolytic dissociation it gives rise to *two* new individuals, the ions. That is, owing to the fact of electrolytic dissociation, there are a greater *number* of individuals present in a given volume of solution than we would have expected. But each individual contributes its own share to the total osmotic pressure, for we have seen that osmotic pressure depends upon the concentration—*i.e.*, upon the number of solute individuals in a given volume, and not upon their chemical nature. Hence, since electrolytic dissociation has brought into existence more individuals than could have been produced from a non-electrolyte at the same grammolecular concentration, it follows that the osmotic pressure of an electrolyte must be greater than that of a non-electrolyte, even though we dissolve equimolecular quantities (say one grammolecule of each weighed out in the solid form) in equal volumes of water. Thus a very dilute solution of KCl possesses twice as many solute individuals as a solution of cane sugar at the same molecular concentration—as indicated by the weighings —because of the fact that, at great dilution, KCl is almost completely dissociated, and each molecule of KCl gives rise to two ions. *The theory of electrolytic dissociation accounts, therefore, for the abnormally high osmotic pressure values exhibited by electrolytes.*

We may make the above statement a little more quantitative in the following way. Suppose that we weigh out one grammolecule of KCl and dissolve it in one litre of water. Suppose that a part or fraction a of this grammolecule is transformed into ions. Then the amount of undissociated KCl is $(1 - a)$ grammolecules, whilst there are likewise a gramions of potassium and a gram-ions of chlorine. Hence the total number of individuals is now $(1 - a) + a + a$, or, in all, $(1 + a)$ individuals. If no dissociation had occurred, a would have been zero. Hence, the ratio of the *actual* number of individuals of solute present to the number which would have been there, had there been no dissociation, is $\dfrac{(1 + a)}{1}$, or simply $(1 + a)$. Hence the osmotic pressure of a KCl solution should be greater than that given by an equimolecular solution of sugar (or other non-electrolyte) in the ratio of $(1 + a)$ to 1. Hence from a measurement of the osmotic pressure we can tell the degree of electrolytic dissociation of the solute. As a matter of fact, this is not the most convenient means of determining a. A more convenient way is to measure the electrolytic conductivity of the solution, from which, as Arrhenius has shown, the degree of dissociation can be obtained in a very accurate manner.

Now the degree of electrolytic dissociation — attained practically instantaneously — remains unchanged with time. It corresponds therefore, to the equilibrium point of a reaction. We may, therefore, apply the law of mass action to find out whether the process of electrolytic dissociation obeys the law or not. Thus, suppose that we have one grammolecule of salt dissolved in V litres of water. Suppose the equilibrium position corresponds to the degree of dissociation a. Then there are $(1 - a)$ grammolecules of undissociated salt in V litres of solution, or the concentration of the undissociated molecules is $\frac{1 - a}{V}$ grammolecules per litre, the concentration of each of the two kinds of ions being $\frac{a}{V}$ gram-ions per litre.

Now, we have seen that the equilibrium constant, or, as it is called in the present case, the *dissociation constant*, K, is simply the product of the concentrations of the ions divided by the concentration of undissociated molecules in the equilibrium state—a state which is reached practically instantaneously in cases of electrolytic dissociation. That is, K is given by the law of mass action in the following form :—

$$K = \frac{\text{concentration of cations} \times \text{concentration of anions}}{\text{concentration of undissociated molecules.}}$$

$$= \frac{\dfrac{a}{V} \times \dfrac{a}{V}}{\dfrac{(1 - a)}{V}}$$

$$= \frac{a^2}{(1 - a)V.}$$

This expression—which is simply the law of mass action applied to the case of electrolytic dissociation—was first stated by Ostwald, and has been tested in numerous cases. We quote below the data obtained in the case of the electrolytic dissociation of the organic acid, acetic acid, which dissociates according to the scheme—

Acetic acid molecule→acetic-anion + hydrogen ion
$$CH_3COOH \rightarrow CH_3COO^- + H^+.$$

We naturally test the applicability of the mass action expression by varying the dilution, V, and measuring the corresponding variation in a, finally inserting the values in the above equation and calculating K.

The values obtained for K are as follows :—

TEMPERATURE 25° CENTIGRADE.

V = volume of solution, in litres, which contains one gram-molecule of acetic acid.	a = degree of electrolytic dissociation.	K
16	0.01673	0.0000179
32	0.02380	0.0000182
64	0.0333	0.0000179
128	0.0468	0.0000179
256	0.0656	0.0000180
512	0.0914	0.0000180
1024	0.1266	0.0000177
		mean = 0.000018

Obviously the law of mass action applies very satisfactorily in this case, as is shown by the constancy in the values for K. Similar results have been obtained in the case of ammonia, and, indeed, several hundred such acids and bases have been examined, and found to dissociate electrolytically in accordance with the principle of mass action. All these acids and bases, to which reference has just been made, are characterised by the fact that although they are electrolytes they are at the same time *weak* electrolytes—*i.e.*, they conduct the current badly. This is due to the fact that they give rise to very few ions. Thus in the case of acetic acid, even when the dilution is·considerable (say V = 1,024 litres), the degree of dissociation is only 0.126—*i.e.*, 12.6 per cent. of complete dissociation.

Quite different behaviour is exhibited by *strong* electrolytes—acids and bases, such as

$$HCl, \ HBr, \ HI, \ HNO_3, \ H_2SO_4, \ NaOH,$$
$$KOH, \ Ba(OH)_2,$$

and salts such as those already mentioned. Strong electrolytes are characterised by the fact that they conduct electricity well, because they are largely ionised, or electrolytically dissociated, the degree of dissociation a at $V = 1,024$ being nearly unity — *i.e.*, nearly complete. These strong electrolytes do *not* obey the law of mass action, the value of K (calculated in exactly the same manner as in the case of acetic acid) showing a large variation with the dilution. In fact, the value of K diminishes steadily as V is increased. In other words, at the greater concentrations of the electrolyte its degree of dissociation a is too great. The cause of this so-called anomaly of strong electrolytes has not yet been discovered, nor has a satisfactory explanation of it been put forward, in spite of much investigation of the problem.

Passing from the particular question of the applicability of the law of mass action to electrolytic dissociation, mention may be made of one very familiar process which involves reactions between ions—*viz.*, the act of neutralisation of a strong acid, such as HCl, with a strong base such as NaOH. On the older view the reaction of neutralisation was regarded as occurring thus—

$$HCl + NaOH = NaCl + H_2O.$$
$$acid + base = salt + water.$$

On the basis of the theory of electrolytic dissociation, however, the HCl, NaOH, and NaCl, being known to be strong electrolytes, must be largely dissociated in aqueous solution into their ions. That is, the process of neutralisation of HCl by NaOH is really—

$$H^+ + Cl^- + Na^+ + OH^- = Na^+ + Cl^- + H_2O.$$

Pure water is known to be almost a non-electrolyte, and is, therefore, in the form of undissociated molecules. On examining the above equation, we see that there are certain terms common to both sides, the Na^+ and the Cl^- terms. They may, therefore, be neglected, for they play no part in the process, the essential reaction in the act of neutralisation being—

$$H^+ + OH^- = H_2O.$$

That is, the act of neutralisation of any strong acid by a strong base is always the same, being, in fact, simply the union of hydrogen ion and hydroxyl ion to form a molecule of undissociated water. Now it is known that the neutralisation of a strong acid by a strong base is accompanied by a marked evolution of heat, the temperature of the solution rising considerably. If, therefore, the process is essentially the same in all these cases—as the theory of electrolytic dissociation requires— then the heat effect should be the same for equimolecular quantities neutralised. This conclusion is in excellent agreement with experiment, as the following table shows.

The value of the heat evolved per grammolecule of water formed is expressed in the heat unit, the calorie. The meaning of the calorie is given in the following chapter.

NEUTRALISATION PROCESS : OBSERVED HEAT EFFECT.

Acid and Base employed.	Heat evolved per Grammolecule of Water formed.
HCl +NaOH	13,700 calories.
HBr +NaOH	13,700 ,,
HI +NaOH	13,700 ,,
HNO_3+NaOH	13,700 ,,
HCl +LiOH	13,800 ,,
HCl +KOH	13,700 ,,
HCl +$\frac{1}{2}$Ba(OH)$_2$	13,900 ,,

Obviously the theory of electrolytic dissociation affords a satisfactory explanation of the fact that the observed heat evolved is practically the same in the neutralisation of *any* strong acid with a strong base. Conversely, we conclude that the heat *absorbed*, when one grammolecule of water ionises into its ions,—*i.e.*, when the reaction

$$H_2O \rightarrow H^+ + OH^-$$

occurs—must amount to 13,700 calories.

In considering any case of electrolytic dissociation the question naturally arises as to the cause of the phenomenon. It is evidently due to the presence of the solvent, for one and the same salt is known to be dissociated electrolytically to different extents in different solvents. Water, indeed, is one of the best solvents for dissociating a solute, provided the solute can be dissociated at all. In the case of a salt such as KCl, it is now generally supposed that the undissociated molecule really consists of two oppositely charged atoms held together by lines of electrical force. When there is no solvent these lines of force start from one of the atoms, and *all* end on the other, both atoms being firmly attached together as a consequence. When, however, the molecule is dissolved in a medium, such as water, it comes under the influence of neighbouring solvent molecules, which detach some of the lines of force drawing them to themselves. The bond of union between the original atoms is thus weakened. In fact, all the lines of force which initially existed between the chlorine and the potassium in the KCl molecule become attached to molecules of water, and the charged atoms become free of one another. The result is that hydrated ions—*i.e.*, ions with their lines of force ending on neighbouring water molecules—are brought into existence, the process being practically instantaneous. The greater the attraction of the molecules of the solvent for the lines of force between the atoms of the solute, the greater the dissociating power of the solvent. Naturally the number of these lines of force depends upon the chemical structure of the molecule of the solute, and, consequently, different substances are dissociated to different extents by a given solvent. This is the cause of the difference between strong and weak electrolytes—weak electrolytes being hard to dissociate, whilst the so-called strong electrolytes are easily dissociated.

There are, of course, numerous applications of the theory of electrolytic dissociation, but these we are unable to discuss further at present.

In all the processes of ionisation and the union of ions which we have been considering, we are dealing with reactions which proceed far too rapidly to be measured. That is, we only deal with equilibrium states. There are, however, several reactions which only partly involve ions, which can occur in solutions at a measurable speed or velocity, for which, therefore, we can calculate the velocity constant. We shall consider two such reactions.

The first of these is the saponification of ethyl acetate by caustic soda. The reaction is—

Ethyl acetate + caustic soda = sodium acetate + ethyl alcohol

$$CH_3COOC_2H_5 + NaOH = CH_3COONa + C_2H_5OH.$$

This reaction occurs in aqueous solution at a measurable speed at ordinary temperatures. The reaction may be followed by titrating separate portions of the solution at various intervals of time. The ethyl acetate and the caustic soda are both solutes, being present in relatively small concentration compared with the concentration of the water. As the reaction proceeds, the concentration of both reacting substances, the ethyl acetate and the caustic alkali, diminishes. It is, therefore, a bimolecular reaction, and we can apply the ordinary bimolecular formula for the rate; *viz.*—

Rate of saponification $= k_{bi} \times C_{\text{ethyl acetate}} \times C_{\text{caustic}}$,

where k_{bi} is the bimolecular velocity constant.

Experiment has shown that this reaction is indeed bimolecular, a good constant value being obtained for k_{bi} which is independent of the time or progress of the reaction.

The second reaction which we shall consider is the inversion of cane sugar. This reaction consists in the transformation of cane sugar into a mixture of dextrose and lævulose by the interaction of water, thus—

cane sugar + water = dextrose + lævulose

$$C_{12}H_{22}O_{11} + H_2O = C_6H_{12}O_6 + C_6H_{12}O_6.$$

Cane sugar, dextrose, and lævulose are optically active, and consequently the reaction can be conveniently followed by means of the polarimeter. Strictly speaking, the reaction between water and cane sugar is bimolecular. Thus we write—

Rate of inversion of cane sugar =
$$k_{bi} \times C_{\text{cane sugar}} \times C_{H_2O}$$
where k_{bi} is the bimolecular velocity constant.

But in dilute aqueous solution the concentration of the water is practically constant—*i.e.*, the term C_{H_2O} is a constant, and it may be merged into the velocity constant itself. We obtain, therefore, an apparently unimolecular process expressed by—
$$\text{Rate of inversion} = k\, C_{\text{sugar}},$$
where k is the velocity constant of a unimolecular process. That this really represents what happens is borne out by experiment, for a good unimolecular velocity constant is actually given by this reaction. This is simply a particular case of a general conclusion—*viz.*, that when one of the reacting substances is present in such large excess (the water in the above case) that its concentration is practically constant throughout the progress of the reaction, then the rate simply depends upon the concentration of the other substance, the concentration of which sensibly diminishes during the course of the reaction.

As in the case of reactions in gases, the velocity constants and equilibrium constants of reactions in solutions vary with the temperature, and, indeed, to much the same extent as in gases.

CHAPTER VI.

HEATS OF REACTIONS.

In the previous chapters dealing with chemical reactions we have fixed our attention upon the material changes involved—*i.e.*, upon the disappearance of one kind of molecule and the formation of another kind. Besides material changes we must also glance briefly at the concomitant energy changes, for every material change is accompanied by an energy change, either small or great. It may be said that, in general, every chemical process is aecompanied by a heat effect—*i.e.*, heat is either evolved or absorbed.

We have already pointed out that heat is one of the forms of energy. If the temperature of a system rises as a result of the chemical change, we say that heat has been generated in the system—*i.e.*, that heat is evolved in the process. If the temperature of a system falls as a consequence of the chemical change, heat has been absorbed by the process.

It is necessary, of course, to express a heat effect quantitatively—*i.e.*, in units of some kind. The usual unit is called the calorie. The calorie is defined as that amount of heat which we must add to one gram of water to raise its temperature one degree—between 15° and 16° centigrade. We make use of this unit in measuring the heat of a chemical reaction. To make a measurement of the heat effect of a reaction we cause the reaction to proceed in a vessel which is immersed in water. We note the temperature of the water at the commencement and at the end of the reaction. Let us suppose that the observed rise in temperature of the water in the outer vessel—called the calorimeter—is 10°, and that the mass of water is 50 grams. Then the heat which has been evolved by the reaction has been sufficient to raise the temperature of 50 grams of water by 10°. That is, the heat evolved in the process is 50×10, or 500 calories. (Due precautions must be taken against accidental losses of heat due to radiation from the vessel.) Further, it is usual to express the heat of a reaction in terms of one grammolecule of the reacting substance or substances which disappear in the process. Thus, suppose that one gram of the reacting substance has, by its chemical change, evolved 500 calories: then, if the molecular weight be M grams, it follows that the heat of the reaction per grammolecule of the substance transformed is 500 M calories. By means of such measurements it has been shown that when one grammolecule of water, (H_2O), is formed by the neutralisation of a strong acid by a strong base—the case cited in the preceding chapter—the heat evolved is 13,700 calories. That section of physical chemistry which deals with the determination of the heat effects of different reactions is called *Thermo-chemistry.*

There is a very fundamental law or principle in thermo-chemistry which is of great utility in enabling us to obtain values for heat effects indirectly—*i.e.*, by calculation from the observed heat effects of other reactions. This principle is known as the law of heat summation. It is really a special case of a more general principle known as the *Principle of the Conservation of Energy.* This principle states that energy cannot be destroyed or generated *de novo ;* it can only be transformed from one form to another. Thus mechanical energy can be transformed into heat energy. For example, when a rapidly-moving body is suddenly stopped by striking against an obstruction, its kinetic energy of motion has obviously disappeared, but it is found that heat has been generated—*i.e.*, the temperature of the " obstruction " has risen. A very familiar case is that of simply rubbing two pieces of matter together. They warm up—*i.e.*, their temperature rises, or heat has been added to them. This heat represents a transformation of a certain part of the mechanical energy of rubbing. In fact, a very careful experiment, carried out many years ago by Joule, in which water was warmed by rotating a paddle vigorously in it, permits of a determination of what is called the mechanical equivalent of heat. The mechanical work done by the paddle was easily measured in ergs. From the rise in temperature of the water the number of calories added to the water was likewise obtained, and these two energy terms must be equal by the principle of conservation of energy. In this way Joule found that one calorie was equivalent to 42 million ergs.

This idea of the indestructibility of energy has been applied to chemical and physical changes in the following way, dealing with heat effects themselves. Let us suppose that a substance is initially in a state A, and that it passes into a new state B, heat being evolved, say q_1 calories. Let the state B then pass into the state C, the change from B to C being accompanied by an evolution of heat q_2. The total heat of the change from A to C *directly* should, therefore, be the same as the heat effect from A to B plus that from B to C. That is, if the observed heat evolved in passing at one step from state A to state C is q_3 calories, then the principle of heat summation states that—

$$q_3 = q_1 + q_2.$$

This additivity in a series of heat effects has been verified experimentally. Let us take an actual physical case. Let us start with one gram of ice and melt it—*i.e.*, take it from the ice state (state A) to the liquid state (state B). It is found that heat is *absorbed* in this process—*viz.*, 80 calories per gram of ice melted. If now the water be vaporised—*i.e.*, if the water state (B) be changed to the steam or vapour state (C)—a further quantity of heat —*viz.*, 530 calories—is absorbed. It follows, therefore, by applying the principle of heat summation, that the heat absorbed in the *sublimation* of ice — *i.e.*, the direct passage from ice to vapour (state A to state C)— must be accompanied by an absorption of heat amounting to 530 + 80, or 610 calories per gram of ice. Many analogous cases are known.

Now let us consider a chemical reaction. On heating solid carbon with oxygen gas the gas carbon dioxide is formed. The chemical equation expressing this is—

$$C+O_2=CO_2.$$

This reaction is accompanied by a large evolution of heat—*viz.*, 94,310 calories for one grammolecule of CO_2 produced. This additional information is expressed by the so-called thermo-chemical equation—

$$C+O_2=CO_2+94,310 \text{ (calories)},$$

where the plus sign denotes that the reaction is accompanied by an evolution of heat. Had the process involved an absorption of heat, a negative sign would have been employed. Further, it is possible to cause carbon monoxide gas to react with oxygen gas to give carbon dioxide gas, the heat evolved in this process being 68,000 calories per grammolecule of CO_2 formed. The thermo-chemical equation is—

$$CO+O=CO_2+68,000 ;$$

or, using double quantities—

$$2CO+O_2=2CO_2+136,000.$$

From the above data it is possible to calculate the heat effect which accompanies the reduction of carbon dioxide to carbon monoxide by carbon, by making use of the general principle already stated.

The reaction the heat of which we wish to measure is—

$$CO_2+C=2CO+x,$$

where x stands for the unknown heat effect. Obviously the value of x may be calculated by simply writing down the thermo-chemical equations of the two reactions whose heat effects we do know and subtracting them in the ordinary algebraic way, thus—

$$2CO+O_2=2CO_2+136,000$$
$$C+O_2=CO_2+94,310$$

whence—

$$2CO+O_2-C-O_2=2CO_2-CO_2+41,690 ;$$
$$\text{or, } 2CO-C=CO_2+41,690 ;$$
$$\text{or, } CO_2+C=2CO-41,690.$$

That is, when carbon dioxide is reduced by carbon to carbon monoxide, heat is *absorbed*, amounting to 41,690 calories per grammolecule of carbon dioxide reduced, or per two grammolecules of CO formed. This mode of calculation is very general, and is of the greatest importance, as it frequently permits us to calculate the heat effect of a chemical reaction which it would be exceedingly difficult, if not impossible, to determine by direct experiment.

There is another way of looking at the heat effect of a reaction which indicates the importance of such determinations. For the sake of simplicity, let us think of the simplest possible type of chemical reaction, *viz.*,—

$$A \longrightarrow B + x \text{ calories.}$$

Let us suppose that the change takes place in an enclosed vessel placed in a calorimeter, and suppose that x calories are evolved per gram-molecule of B formed. It is clear that this heat energy must have come from the chemical system · itself. That is, it has its origin in the internal energy of the molecules of the system. In fact, the heat effect which is observed is simply the difference between the average internal energy of the substance A and the internal energy of the substance B, the mass of each considered being one gram-molecule. Heat determinations, therefore, allow us to estimate the *difference* between the internal energy or so-called energy content of the substances involved. Substances with different chemical constitution possess different energy contents, and, in fact, the energy content is as important a quantity as the constitution itself.

Such considerations of energy changes have led to considerable advances in knowledge; but it would take us beyond the scope of this book to pursue the subject further. Enough has been said, however, to indicate the importance of determining such energy changes as well as the material changes which occur in a chemical reaction, for by such knowledge we are enabled to obtain further insight into the fundamental problem of the ultimate constitution and behaviour of matter.

INDEX.

ANIONS, 104.
Anode, 104.
Anomaly of strong electrolytes, 112.
Atomic nucleus, 73.
Atoms, structure of, 67 *seq.*
Attraction between molecules, 51 *seq.*; 56.
Avogadro hypothesis, 38 *seq.*
—— ——, kinetic basis of, 47, 48.
Avogadro number, 58 *seq.*

BIMOLECULAR reactions, 87, 116 *seq.*
Boyle's law, 30 *seq.*
—— ——, kinetic basis of, 40 *seq.*
Brownian movement, 60 *seq.*

CALORIE, 120.
Cathode, 104.
Cations, 104.
Chemical equilibrium, 77 *seq.*
Concentration, 81.
Conservation of energy, principle of, 121.
Critical state, 54 *seq.*

DEGREE of electrolytic dissociation, 106, 108.
Dissociation constant (of an electrolyte), 109–111.
Dyne, 17.

ELECTROLYTES, 103 *seq.*
Electrolytic dissociation theory, 105 *seq.*
Electron, 67 *seq.*
——, charge on, 69.
——, mass of, 69.
Equilibrium, chemical, 77 *seq.*, 91.
—— constant, 85, 109 *seq.*
Erg, 24.

GASES, 30 *seq.*
——, chemical reactions in, 76 *seq.*
Gases, compressed, 49 *seq.*
Gas law, 37, 38.
Gas "constant," 37.
Gay-Lussac's law, 33 *seq.*
—— —— ——, kinetic basis of, 46, 47.

HEAT of reaction, 119 *seq.*
Heat summation, principle of, 122 *seq.*

INTERNAL energy, 125.

KINETIC energy, 14.
—— —— of molecules, 43 *seq.*

LIQUIDS, 49.

MASS action, principle of, 79.

127

Molecular magnitudes, 58 *seq.*
Molecules, diameter of, 63.
——, existence of, 58 *seq.*
——, mean free path of, 64.
——, velocity of, 65.
——, volume of, 51, 63.

NEUTRALISATION, heat of, 114.
——, mechanism of, 112 *seq.*

OSMOTIC pressure, 98 *seq.*

PERIODIC law, 72.
Physical changes, 27 *seq.*
Physical chemistry, definition of, 9 *seq.*
Polymerisation of water, 28.
Potential energy, 14.
Principle of mass action, 79 *seq.*

RATE of a reaction, 76, 81.
Reaction velocity in solutions, 116 *seq.*
Reversible reactions, 77.

SEMI-PERMEABLE membrane, 99.
Solid state, 56.
Solute, 97.
Solutions, reactions in, 97 *seq.*, 116 *seq.*
Speed of a reaction, 76, 81.

TEMPERATURE, absolute, 36, 44.
——, kinetic basis of, 43.
—— on reaction velocity, effect of, 96.
Thermo-chemistry, 121.

UNIMOLECULAR reaction, 82 *seq.*

VAN DER WAALS' equation, 52 *seq.*
Van 't Hoff's osmotic law, 102.
Velocity constant, 82 *seq.*
Velocity of a reaction, 76, 81.

THE END.

PRINTED IN GREAT BRITAIN AT
THE PRESS OF THE PUBLISHERS.

University of Toronto Library

DO NOT

REMOVE

THE

CARD

FROM

THIS

POCKET

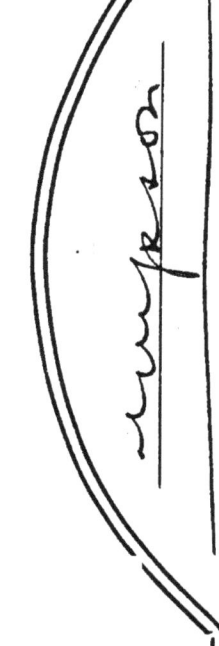

Lightning Source UK Ltd.
Milton Keynes UK
UKHW02f2019220818
327653UK00014B/1355/P